Fluoride in Drinking Water

Status, Issues, and Solutions

Fluoride in Drinking Water

Status, Issues, and Solutions

A.K. Gupta
S. Ayoob

CRC Press is an imprint of the
Taylor & Francis Group, an **informa** business

CRC Press
Taylor & Francis Group
6000 Broken Sound Parkway NW, Suite 300
Boca Raton, FL 33487-2742

First issued in paperback 2023

Version Date: 20160322

ISBN-13: 978-1-4987-5652-5 (hbk)
ISBN-13: 978-1-138-32266-0 (pbk)
ISBN-13: 978-0-429-09175-9 (ebk)

DOI: 10.1201/b21385

Library of Congress Cataloging-in-Publication Data

Names: Gupta, A. K. (Ashok Kumar) author. | Ayoob, S., author.
Title: Fluoride in drinking water : status, issues and solutions / A.K. Gupta and S. Ayoob.
Description: Boca Raton : Taylor & Francis, a CRC title, part of the Taylor & Francis imprint, a member of the Taylor & Francis Group, the academic division of T&F Informa, plc, [2016] | Includes bibliographical references and index.
Identifiers: LCCN 2016000273 | ISBN 9781498756525 (hardcover : alk. paper)
Subjects: LCSH: Fluorides--Environmental aspects. | Fluorides--Toxicology. | Water--Purification. | Water--Fluoridation--Environmental aspects. | Water--Fluoridation--Health aspects.
Classification: LCC TD427.F54 G87 2016 | DDC 363.17/91--dc23
LC record available at http://lccn.loc.gov/2016000273

Visit the Taylor & Francis Web site at
http://www.taylorandfrancis.com

and the CRC Press Web site at
http://www.crcpress.com

Contents

Preface ... xi
Acknowledgments ... xiii
Authors ... xv

1. Fluoride in Drinking Water: A Global Perspective ... 1
1.1 Introduction ... 1
1.2 Drinking-Water Scenario ... 1
1.3 Geogenic Pollutants ... 3
1.3.1 Fluorine: The Chemical Profile ... 3
1.3.2 Sources of Fluoride ... 3
1.4 Fluoride in Humans ... 5
1.5 Genesis of Fluoride in Groundwater ... 5
1.6 Summary ... 8
References ... 8

2. Scenario of Fluoride Pollution ... 11
2.1 Introduction ... 11
2.2 Global Scenario ... 11
2.2.1 Asian and African Scenario ... 11
2.2.2 Indian Scenario ... 20
2.3 Summary ... 21
References ... 22

3. Dental Fluorosis ... 27
3.1 Introduction ... 27
3.2 Dental Effects of Fluoride ... 27
3.2.1 Dental Caries (Tooth Decay) ... 28
3.2.2 Prevention of Dental Caries by Fluoride ... 29
3.2.3 Role of Fluoride in Dental Decay ... 29
3.3 Dental Fluorosis: History and Occurrence ... 30
3.4 Development of Dental Fluorosis ... 31
3.4.1 Physical Symptoms of Dental Fluorosis ... 32
3.4.2 Issues of Dental Fluorosis ... 33
3.4.3 Prevalence of Dental Fluorosis ... 33
3.5 Summary ... 35
References ... 36

4. Skeletal Fluorosis 39
4.1 Introduction 39
4.2 Action of Fluoride on Bone 39
4.3 Fluoride Exposure Level and Skeletal Fracture 40
4.4 Skeletal Fluorosis 41
4.4.1 Crippling Skeletal Fluorosis 43
4.5 Fluoride Level and Effects Related to Skeletal Fluorosis 44
4.6 Significance of Other Factors 45
4.7 Recent Developments 46
4.8 Summary 47
References 48

5. Stress Effects of Fluoride on Humans 51
5.1 Introduction 51
5.2 Nonskeletal Fluorosis 51
5.3 Fluoride and Cancer 51
5.4 Fluoride and Gastrointestinal System 53
5.5 Other Health Effects 54
5.6 Summary 55
References 56

6. Fluoride in the Environment and Its Toxicological Effects 59
6.1 Introduction 59
6.2 Sources of Environmental Exposure 59
6.3 Environmental Transport, Distribution, and Transformation 60
6.4 Environmental Levels and Human Exposure 61
6.4.1 Fluoride from Dental Products 62
6.4.2 Fluoride from Food and Beverage 63
6.4.3 Fluoride in Soil 68
6.4.4 Fluoride in Tobacco and Pan Masala 69
6.4.5 Fluoride from Occupational Exposure 69
6.4.6 TF Exposure 71
6.5 Effects of Fluoride on Laboratory Animals and In Vitro Systems 71
6.6 Effect of Fluoride on Aquatic Organisms 76
6.7 Effect of Fluoride on Plants 77
6.8 Effect of Fluoride on Animals 78
6.9 Guidelines Values and Standards 79
6.10 Summary 80
References 81

7. Defluoridation Techniques: An Overview 87
7.1 Introduction 87
7.2 Coagulation 87
7.2.1 Lime 88

7.2.2 Magnesium Oxide 88
7.2.3 Calcium and Phosphate Compounds 89
7.3 Co Precipitation of Fluoride 90
7.3.1 Alum 90
7.3.2 Alum and Lime (Nalgonda Technique) 91
7.4 Adsorption 93
7.4.1 Bone and Bone Charcoal 93
7.4.2 Clays and Soils 94
7.4.3 Carbonaceous and Other Adsorbents 96
7.4.4 Alumina 99
7.4.5 Activated Alumina 101
7.4.6 Other Alumina-Based Adsorbents 103
7.5 Electrochemical Methods 106
7.5.1 Electrocoagulation 106
7.5.2 Electrosorption 109
7.6 Membrane Processes 109
7.6.1 Reverse Osmosis 110
7.6.2 Nanofiltration 112
7.6.3 Electrodialysis 113
7.7 Defluoridation Techniques: A Summary 114
7.8 Summary 115
References 117

8. Adsorptive Removal of Fluoride: A Case Study 123
8.1 Introduction 123
8.2 Materials and Methods 123
8.2.1 Reagents and Adsorbate 123
8.2.2 Synthesis of the Adsorbent 124
8.2.3 Instrumentation 124
8.2.4 Characterization of the Adsorbent 124
8.3 Batch Studies 125
8.3.1 Effect of Process Parameters 125
8.3.2 Equilibrium Studies 126
8.3.3 Column Studies 127
8.4 Theoretical and Mathematical Formulations 128
8.4.1 Adsorption Capacity 128
8.4.2 Kinetic Modeling 129
8.4.2.1 Pseudo-First-Order Model 129
8.4.2.2 Pseudo-Second-Order Model 130
8.4.2.3 Intra Particle Diffusion Model 131
8.4.2.4 Elovich Equation 131
8.4.2.5 Arrhenius Equation 132
8.4.3 Elucidation of Rate-Limiting Step 132
8.4.4 Adsorption Equilibrium and Isotherms 133
8.4.4.1 Langmuir Isotherm 134

8.4.4.2 Freundlich Isotherm ... 135
8.4.4.3 Dubinin–Radushkevich (D–R) Isotherm ... 136
8.4.4.4 Selection of Best-Fitting Isotherm ... 137
8.4.4.5 Natural and Synthetic Systems ... 137
8.4.4.6 Concentration and Dose Variation Studies ... 138
8.4.5 Factors Influencing Adsorption ... 138
8.4.5.1 Adsorbent Dose ... 138
8.4.5.2 Contact Time ... 139
8.4.5.3 Agitation Rate ... 139
8.4.5.4 Effect of pH and Coexisting Ions ... 139
8.4.5.5 Temperature ... 140
8.4.5.6 Ionic Strength ... 141
8.4.6 Behavior of Adsorption Columns ... 141
8.4.7 Analysis and Modeling of Breakthrough Profile ... 143
8.4.7.1 Hutchins BDST Model ... 143
8.4.7.2 Thomas Model ... 144
8.4.7.3 Yoon–Nelson Model ... 145
8.4.7.4 Clark Model ... 146
8.4.7.5 Wolborska Model ... 146
8.4.7.6 Bohart and Adams Model ... 147
8.4.8 Regeneration ... 147
8.5 Results and Discussions ... 148
8.5.1 Characterization of the Adsorbent ... 148
8.5.2 Kinetics Studies ... 149
8.5.2.1 Agitation Rate ... 149
8.5.2.2 Adsorbent Dosage ... 150
8.5.3 Kinetic Profile of Fluoride Uptake ... 152
8.5.3.1 Pseudo-First-Order Model ... 153
8.5.3.2 Pseudo-Second-Order Model ... 153
8.5.3.3 Intra Particle Surface Diffusion Model ... 154
8.5.3.4 Elovich Model ... 154
8.5.3.5 Arrhenius Equation ... 155
8.5.4 Elucidation of Rate-Limiting Step ... 156
8.5.5 Fluoride Removal Mechanism ... 158
8.5.6 Isotherm Studies ... 162
8.5.6.1 Effects of Temperature ... 163
8.5.7 Performance Evaluation of ALC in Natural and Synthetic Systems ... 163
8.5.7.1 Effect of pH, Ionic Strength, and Temperature ... 164
8.5.7.2 Effects of Other Ions ... 166
8.5.8 Column Studies ... 167
8.5.8.1 Effect of Process Parameters on Breakthrough ... 168

8.5.9 Application of Sorption Models 171
8.5.9.1 Comparison of the Applied Models (Synthetic Water) 175
8.5.9.2 Comparison of the Applied Models (Natural Water) 176
8.5.10 Fluoride Desorption Studies 177
8.6 Summary of the Case Study 179
8.7 Conclusions of the Case Study 184
References 185

Index 189

Preface

Of late, geogenic pollutants have been responsible for polluting groundwater. This has resulted in the pollution of groundwater evolving as a distinct academic discipline in the arena of environmental science and engineering. So we felt the scope of books related to environmental engineering needs to be inclusive of a broader range of such key environmental issues of our time. The genesis of this book stems from such a feeling.

Textbooks on environmental engineering are generally confined to traditional isolated subjects that are based on various courses and syllabi. Given the impending issues related to fluoride in drinking water and associated human health issues, we believe it is pedagogically undesirable to keep the issues of fluoride outside our syllabi and classrooms.

The main audience of this book is expected to be the research community, faculties, scholars, students (undergraduate, graduate, and postgraduate), and technical professionals who are learning or working in the arena of environmental science and engineering and dealing with drinking water quality, supply, and management. For practicing environmental and water supply, engineers, and professionals working in related areas, this book is tailored as a guide for providing a solid fundamental understanding of the crux of global fluoride pollution. For anyone who needs a substantial overview of the present global water quality and fluoride-related issues, this book provides a strong, specific, and updated database.

We have structured the chapters with continuity to get a clear vision and understanding of the magnitude and gravity of the problem. This book is an illustration of a research effort that is attempted to facilitate a deeper understanding of the issue of fluorosis. To have an in-depth understanding of any related issues, readers are advised to refer to the advanced literature cited in our publications. It is our hope this book may serve as an inspiration in the journey of research to a world of fluoride-free drinking water, which we believe is not a distant dream.

A. K. Gupta and S. Ayoob

Acknowledgments

We remember the love and prayers from our parents and teachers, without which our lives would not have been as they are now. We appreciate our family members for preserving a cheerful home and our scholarly friends and students for their lovable support.

Authors

A.K. Gupta earned a PhD in environmental science and engineering at the Indian Institute of Technology Bombay, India. Currently, he is a professor in the Environmental Engineering Division of the Civil Engineering Department at the Indian Institute of Technology Kharagpur, India, and he is actively involved in teaching, research, and consultancy. His research interests are focused on water treatment, environmental impact assessment, monitoring, and modeling of air and water pollution, geogenic pollutant scavenging, and so on. He has more than 60 publications in top-ranking international journals. He is a renowned technical consultant in the arena of environmental engineering and has more than 30 completed/ongoing projects of national and international importance to his credit.

S. Ayoob is a professor in the Department of Civil Engineering and the principal of TKM College of Engineering, Kerala, India. He graduated in civil engineering from the TKM college of engineering, Kollam, India, and earned master's and doctoral degrees at the Indian Institute of Technology Kharagpur, India. He also served as course leader and headed the Department of Health, Safety, and Environmental Management at the International College of Engineering and Management, the Sultanate of Oman (affiliated with the University of Central Lancashire, United Kingdom) for a short period.

1

Fluoride in Drinking Water: A Global Perspective

1.1 Introduction

Water is life as it aids in nurturing the lives of all biota. The availability of clean water has become obliquely central to the quality of human life. However, there are deep currents that affect the water dynamics of today's world. Currently, two-thirds of our planet is covered water; however, the unfortunate paradox is that in the next decade, bulk of the human population will lack access to safe drinking water. This acute shortage of drinking water may change our traditional perceptions about both the quality and usage of water in the future. Further, the presence of geogenic pollutants in groundwater makes this issue more complex. Of late, the presence of fluoride in drinking water has attracted much attention in the scientific world due to the impending issues associated with human health and well-being.

1.2 Drinking-Water Scenario

In 2014, it was reported by the World Health Organization that around 750 million people from the poor and socially deprived sections of society all over the world do not have access to improved drinking-water sources. Almost one-fourth of the people living in rural habitations use and drink untreated surface water. In 2015, it was estimated that around 550 million do not have access to improved drinking-water sources.[1] It is predicted that in 2035, there will be a one-third reduction in the per capita drinking-water availability. By 2025, around 34% of the global population will face acute drinking-water shortage.[1] This sorry state of affairs is reflected in the terribly short supply of good quality water in many parts of the world. As a result, the global water supply system is beleaguered at both the demand and supply ends. The acute scarcity of drinking-water sources and increased competition in the early periods of the twenty-first century, the traditional ways of using and valuing water have taken a dramatic turn. Thus, the scarcity of water and the limited access to safe drinking-water sources are predicted to be the most challenging and crucial environmental issues of the future in preserving and defining the quality of life on earth.[2,3]

Groundwater has been perceived to be the safest of all the drinking-water sources available on the surface of the earth. As a result, half of the global population blindly relies on groundwater sources for both drinking and survival. Apart from this, in many regions of the world, groundwater sources turn out to be the single largest source of supply for drinking. Further, in many communities, these sources appear to be the only economically viable option for drinking, as they supply reliable quality water and stable quantity water compared with water from surface sources. Since groundwater plays such a crucial role in the existence of the majority of the global human population, its availability, safety, and purity have become issues of critical concern for many habitations across the globe.[3,4]

Of late, due to rapid urbanization and industrialization, more xenobiotic substances are getting diffused into different spheres of the earth. A considerable portion of these substances rests with the biosphere, of which water sources occupy a considerable share. Water sources act as "sinks" for many of these pollutants, resulting in drinking-water pollution and water scarcity. This situation is more serious in developing countries, as they are grappling with acute issues related to both scarcity and contamination of drinking water. Intrinsically, the excessive groundwater pumping that is disproportionate to recharge will also lead to the depletion of water, thus posing challenges to drinking-water supply systems. Plenty of examples are available to validate this point. In many developing countries such as India, in urban groundwater sources cater to around one-half of the water requirements. In rural areas, groundwater sources alone cater to more than two-thirds of the total water demand, thus resulting in the situation appearing really critical. The indiscriminate tapping of groundwater created alarmingly low levels of the water table in many parts of the developing world.[3,5,6] Thus, inadequate access to safe drinking water, on the one hand, and its ever-increasing intimidation from abundant contaminants, on the other hand, make the global drinking-water scenario more complex. As a result, the world is heading toward a water crisis. This is a crisis affecting both the quantity and quality of water. In a little more than half a century, this global water crisis has evolved, mainly affecting the developing world in and around the arid and semi arid regions, especially areas where groundwater is the main source of drinking water. The drilling of tube wells for agricultural purposes is often unregulated, though it is supplemented by subsidized electricity for pumping. Much of the progress in food production, such as the Green Revolution in India and other countries, has taken place unsustainably at the expense of groundwater. This has triggered a severe drawdown of groundwater tables[7]; it has also had lasting and often irreversible impacts on groundwater, resulting in a synergy of water quality issues.

1.3 Geogenic Pollutants

Of late, the entry of geogenic pollutants such as fluoride and arsenic into groundwater aquifers has turned out to be a decisive environmental problem the world over. This situation is extremely critical in developing countries such as China and India. In India, fluoride has become endemic in approximately 37,000 habitations, whereas issues of arsenic are diffused into around 3,200 habitations, thus exhibiting the dominance of the issue. The presence of excess fluoride in drinking water raises a red flag of concern, as it initiates fluorosis in various proportions, thereby reducing the quality of human life. It was estimated that people from more than 35 nations across the globe are under the threat of fluoride attack. The number of people affected with the "risk of fluorosis" has probably crossed 200 million.[8–10] Therefore, fluoride in groundwater can be treated as a critical driver in defining the quality of groundwater in many parts of the world. The steady increase in the number of people falling prey to fluoride pollution, especially in the developing world, brings the issue under global focus. Thus, it would be interesting to have a brief overview on the pathways of fluoride into groundwater, the chemical profile and geo-chemistry of fluoride, and the status of global fluoride pollution while narrating its context and relevance.

1.3.1 Fluorine: The Chemical Profile

Fluorine is the ninth element in the periodic table, with an atomic weight of 18.9984. It belongs to the group VII A. It is rated thirteenth in abundance and is estimated to be widely distributed at 0.3 g/kg of the earth's crust. Elemental fluorine is the most electronegative and reactive of all elements; as a result, it rarely occurs naturally in the elemental state. Its electronegative nature demonstrates that it has a strong tendency to acquire a negative charge in solution, forming fluoride ion (F^-). Except inert gases, it can bond with every other element, thus forming stable electronegative bonds. "Fluorine reacts with other elements to produce ionic compounds like hydrogen fluoride and sodium fluoride in water and upon dissociation forms negatively charged fluoride ion."[3,6,11,12]

1.3.2 Sources of Fluoride

Fluoride is an abundant trace element that is found with an average concentration of 625 mg/kg of fluorine in the earth's crust. However, its occurrence is found to vary depending on the types of rocks (from 100 mg/kg in limestones to 2000 mg/kg in volcanic rocks).[13] The rich underlain presence of crystalline igneous and metamorphic rocks in regions of India, Sri Lanka, Senegal, Ghana, and South Africa along with areas of volcanic and associated hydrothermal activity contributes to fluoride. Since "fluorine has a higher affinity for silicate melts than solid phases it is progressively enriched in magmas and hydrothermal solutions with time due to magmatic differentiation."[3] As a result, the hydrothermal vein deposits and rocks that crystallize from highly evolved magmas often contain fluorite-, fluorapatite-, and fluoride-enriched micas and/or amphiboles. Based on the percentages of silica and calcium present in magma, cryolite, villiaumite, and/or topaz can also be formed. The highest fluoride levels were reported from regions that were predominately occupied by crystalline igneous and metamorphic rocks. These rocks are associated with syenites, granites, quartz monzonites, granodiorites, felsic and biotite gneisses, and alkaline volcanic types. It is suggested that the presence of biotite alone may produce dissolved fluoride concentrations in groundwater to a level of more than 4 mg/L.[13]

The parent rock serves as the most natural contributor of fluoride into drinking water. However, fluorite, the only principal mineral of fluorine, is regarded as an accessory mineral in granitic rocks. Granite rocks are reported to exhibit fluoride concentrations of 20–3600 mg/L. Apatite, muscovite, amphibole, hornblende, pegmatite, mica, biotite, villiaumite, and certain types of clays are also found to contain fluorine. The reported natural Indian sources include the following: the hard rock terrains (south of Ganges valley) in the arid north-western part; fluoride-rich rocks and canal-irrigated black cotton soils of Karnataka; the dark mineral fraction of gneisses of Tamil Nadu; granites, minerals such as sepiolite and palygorskite, acid volcanic and basic dikes of Rajasthan; soils and clays of Gujarat; granitic rocks of Andhra Pradesh; tourmaline-bearing pegmatites of Maharashtra; and sodic soils in irrigated areas of Haryana and Andhra Pradesh.[3,14–19] High fluoride concentrations can also result from "anion exchange (OH^- for F^-) on certain clay minerals, weathered micas and oxyhydroxides that are typically found in residual soils and sedimentary deposits."[3] Areas underlain by alkaline volcanic rocks and sedimentary formations that contain fluorapatite- and/or fluoride-enriched clay minerals may contribute to fluoride concentration. Crystalline basement rocks such as felsic intrusive rocks and their metamorphic forms are also prone to fluoride dissolution. Though fluoride is not readily leached from soils due to its strong associations with the soil components (only 5%–10% of the total fluoride in soil is water soluble), its concentration may increase with depth to the tune of 200–300 mg/L.[20] The rate of fluoride dissociation depends on the soil chemistry, chemical form, climate of the region, and deposition rates. It is reported that in acidic soils with a pH less than 6, fluorides can form complexes with iron and aluminum; they can also form a bonding with clay by replacing hydroxide from the clay surface. pH plays a crucial role in this adsorption process as it turns significant at a pH of 3–4, whereas it is reduced at a pH higher than 6.5. The application of fertilizers under intensive irrigation may result in releasing fluoride into groundwater. Alkalinization may enhance the concentration of fluoride content in irrigated lands and soils.[20,21]

1.4 Fluoride in Humans

It was estimated that 99% of the absorbed fluoride in humans gets deposited in bones and teeth. Though fluoride does not get accumulated in most soft tissues, such as hydrogen fluoride (HF), it can find its way into the intracellular fluid of soft tissues. It is plausible that within kidney tubules, fluoride may get concentrated at high levels, even at a higher concentration than plasma. Due to this relatively high fluoride-level exposure, kidneys are regarded as the most vulnerable sinks of fluoride, resulting in them becoming an impending target of acute and chronic fluoride toxicity.[21,22] Since the transportation of fluoride from plasma to milk is minimal, the observed fluoride levels in human milk are only to the tune of 5–10 μg/L.[20] The level of fluoride concentration in saliva reflects the plasma fluoride availability. Only low concentrations of fluoride are reported in sweat (around one-fifth of plasma levels). Renal excretion of fluoride may be 35%–70% of intake in adults. As a result, urine, plasma, or saliva could be used as biomarkers of fluoride exposure. The average per capita dietary intake range of fluoride will be 0.020–0.048 mg/kg for adults (living in regions with fluoride concentrations of 1.0 mg/L in water). Although a "no-observed-adverse-effect level (NOAEL) of 0.15 mg fluoride/kg/day and a lowest observed-adverse-effect level (LOAEL) of 0.25 mg fluoride/kg/day of fluoride in human" are suggested, these levels are still under scientific debate.[3,20,23–25]

1.5 Genesis of Fluoride in Groundwater

The origin of fluoride in groundwater is due to an interaction between groundwater and surface water with rocks containing fluoride-rich mineral. The presence and concentration of fluoride in groundwater are a reflection of the amount of concentration of fluoride-bearing minerals present in their parent rock types. The decomposition and dissolution activities of the rock types that are exhibited through rock–water interactions play a crucial role.[26] In the developing countries such as India, since the contribution from drinking water is the most significant source of fluoride entry into the human body, a thorough understanding of the geo-chemistry of fluoride in groundwater appears relevant. It is observed that the rainwater falling on the earth gets charged by different sources of CO_2 from the soil and the atmosphere, in addition to the biochemical reactions between bacteria and organic matter during its descent. Thus, the rainwater may turn slightly acidic due to the formation of carbonic acid. As a result, during percolation, the secondary salts present in the soil (mixture of varying content of $NaHCO_3$, NaCl, and Na_2SO_4) may get leached out. In phosphate fertilizer-applied lands, soils most likely contain different percentages and proportions of fluoride-rich materials and compounds. Simultaneously, an ion-exchange reaction takes place, with exchangeable cations present in the soil–clay complex as follows:[27]

$$CaX_2 + 2Na^+{}_{(aq)} \leftrightarrow 2NaX + Ca^{2+}{}_{(aq)} \quad (1.1)$$

where X is the clay mineral. As demonstrated by the equations cited, the hydrogen-ion concentration in groundwater is increased due to the dissolution of CO_2. The calcareous minerals, especially $CaCO_3$, also get dissolved as follows:[27–29]

$$CO_2 + H_2O \rightarrow H_2CO_3 \quad (1.2)$$

$$H_2CO_3 \rightarrow H^+ + HCO_3^- \quad (1.3)$$

$$HCO_3^- \rightarrow H^+ + CO_3^{2-} \quad (1.4)$$

$$CaCO_3 + H^+ + 2F^- \rightarrow CaF_2 + HCO_3^- \quad (1.5)$$

$$CaF_2 \rightarrow Ca^{2+} + 2F^- \quad (1.6)$$

The groundwater charged with alkalinity may mobilize F^- from weathered rocks, soils, and CaF_2, resulting in the precipitation of $CaCO_3$ as follows:[29]

$$CaF_2 + 2HCO_3^- \rightarrow CaCO_3 + 2F^- + H_2O + CO_2 \quad (1.7)$$

The dissolutional and assimilating activity of fluoride gets enhanced with the excess presence of sodium bicarbonates in groundwater as follows:[29]

$$CaF_2 + 2NaHCO_3 \rightarrow CaCO_3 + 2Na^+ + 2F^- + H_2O + CO_2 \quad (1.8)$$

The CaF_2 has a solubility product (K_{sp}) as follows:[30]

$$K_{sp} = [F^-]^2 [Ca^{2+}] = 4.0 \times 10^{-11} \quad (1.9)$$

The low-solubility product (Equation 1.9) suggests that high fluoride concentrations in groundwater are generally associated with low calcium content (with a negative correlation between the two ions) and high bicarbonate ions (in some cases with high nitrate ions).[3] Also, it was observed that groundwater is generally undersaturated with respect to fluorite (in some cases, it may be saturated with both calcite and fluorite).[3,27] It has been established that the pH of groundwater plays a significant role in defining the concentration of fluoride in groundwater. An alkaline environment (within a pH range 7.6–8.6) with a high bicarbonate concentration is said to be more conducive for fluoride dissolution. Thus, the weathering of primary minerals in rocks appears to be the main contributing factor of fluoride into groundwater. The weathering of rocks results in the leaching of fluoride-containing minerals into groundwater. Being the most predominant mineral, mainly the presence of fluorite clues in about the concentrations of fluoride in groundwater. As evidenced in Equation 1.9, the low-solubility product is an irrefutable proof that low levels of calcium result in high fluoride concentrations in water. Further, groundwater in the sodium bicarbonate and bicarbonate chloride types exhibits high fluoride concentrations. The ion-exchange mechanism, as suggested in Equation 1.1, turns significant in the context of reported excess fluoride in groundwater in regions of high sodicity of soil. This process and the mechanisms related to it turn relevant in issues pertaining to reported high fluoride concentrations near the major south Indian irrigation projects.[16,17,26,28,31–33] In 2012, the most recent research conducted in the rocks of Hangjinhouqi in a fluoride endemic area of China suggests that due to CaF_2 solubility control, Na-predominant water is favorable for fluoride enrichment with low total dissolved solid (TDS).[34] The elevated fluoride concentrations in the groundwaters of south-eastern Pakistan region (1.13–7.85 mg/L) are also attributed to the enhanced fluorite solubility due to Ca depletion, high ionic strength, and the release of fluoride from colloid surfaces under high pH conditions.[3,35] The excess fluoride concentration in the Kolar and Tumkur districts in Karnataka, India, is also attributed to the reduced levels of calcium-ion concentration in groundwater due to the calcite precipitation.[36] It is also reported that the dry and hot climate enhances fluoride levels in the groundwater nearer to the surface. In regions with high rainfall, the dilution effects in groundwater may outweigh the enrichment effects. In dry regions where precipitation rates are lower than evaporation, the fluoride generated from the dissolution of fluoride-bearing minerals may move toward the surface as a result of evaporation. This causes an increase in fluoride concentrations in groundwater. The hydrolysis (OH^- in water exchanges for F^-) of F-bearing silicates such as muscovites (Equation 1.10) and biotites (Equation 1.11) in alkaline soda water can be expressed as follows:[37,38]

$$KAl_2(AlSi_3O_{10})F_2 + 2OH^- = KAl_2(AlSi_3O_{10})(OH)_2 + 2F^- \quad (1.10)$$

$$KMg_3(AlSi_3O_{10})F_2 + 2OH^- = KMg_3(AlSi_3O_{10})(OH)_2 + 2F^- \quad (1.11)$$

Thus, it could be inferred that the enrichment of fluoride in groundwater results from water–rock interactions of F-bearing silicates. The dissolution of fluorite may get further triggered by its enrichment through evapo-transpiration.

1.6 Summary

- The average per capita water availability may get reduced by one-third over the next two decades. As a result, by 2025, one-third of humanity will be under the risk of severe water scarcity.
- Inadequate access to safe drinking water has become the most crucial challenge to the sustainable water supply systems of the world.
- The world is heading toward a water crisis. This global water crisis mainly affects the developing world in and around the arid and semi arid regions where groundwater is the main source of drinking water.
- The entry of geogenic pollutants such as fluoride and arsenic into groundwater aquifers has become an issue of global concern, especially in developing countries such as India and China.
- Around 200 million people from more than 35 nations the world over are "at risk" of fluorosis.
- The main source of fluoride in soil is obviously the parent rock itself. The origin of fluoride in groundwater is mainly due to the interaction between groundwater and surface water with rocks containing fluoride-rich minerals.
- The rate of fluoride dissociation depends on the chemical form, rate of deposition, soil chemistry, and climate.
- Around 99% of the absorbed fluoride in humans gets deposited in bones and teeth. The fluoride accumulation within kidney tubules may be very high compared with the plasma. As a result, the kidney could be considered an impending sink, site, and target of the acute fluoride toxicity.

References

1. WHO and UNICEF (2014). *World Health Organization and UNICEF, Progress on Sanitation and Drinking-Water–2014 Update.* Geneva, Switzerland: WHO Press.
2. WWC (2003). World Water Council, 3rd World Water Forum, Press Release, Crucial water issues to be addressed. Tokyo, Japan: Secretariat of the 3rd World Water Forum, 2003. http://www.worldwatercouncil.org/download/PR_curtainraiser_10.03.03.pdf.
3. Ayoob, S. and Gupta, A.K. (2006). Fluoride in drinking water: A review on the status and stress effects. *Critical Rev. Environ. Sci. Technol.*, 36, 433–487.
4. WHO and UNICEF (2004). *WHO and UNICEF, Meeting the MDG Drinking Water and Sanitation Target: A Midterm Assessment of Progress.* New York: WHO Geneva and UNICEF, 2004.

5. Maria, A. (2003). The Costs of Water Pollution in India, CERNA, Ecole Nationale Superieure des Mines de Paris, Paris, France. Revised version, Paper Presented at the Conference on Market Development of Water and Waste Technologies through Environmental Economics, 30th–31st October 2003, New Delhi, India.
6. WHO (2004). Fluoride in drinking water, Background document for preparation of WHO Guidelines for drinking water quality. Geneva, Switzerland: World Health Organization, http://www.who.int/water_sanitation_health/dwq/guidelines/en/.
7. Edmunds, W.M. (2009). Geochemistry's vital contribution to solving water resource problems. *Appl. Geochem.*, 24, 1058–1073.
8. MRD (2004). First report: Standing Committee on Rural Development, Ministry of Rural Development (Department of Drinking Water Supply), Presented to the Fourteenth Lok Sabha, Lok Sabha Secretariat, New Delhi, India, p. 33.
9. Daw, R.K. (2004). Experiences with domestic defluoridation in India, Proceedings of the 30th WEDC International Conference on People-Centred Approaches to Water and Environmental Sanitation, Vientiane, Lao PDR, pp. 467–473.
10. Ayoob, S., Gupta, A.K. and Bhat, V.T. (2008). A conceptual overview on sustainable technologies for the defluoridation of drinking water. *Crit. Rev. Environ. Sci. Technol.*, 38, 401–470.
11. Mackay, K.M. and Mackay, R.A. (1989). *Introduction to Modern Inorganic Chemistry*, 4th edn, p. 339. Englewood Cliffs, NJ: Prentice Hall.
12. Cotton, F.A. and Wilkinson, G. (1988). *Advanced Inorganic Chemistry*, p. 546. New York: John Wiley and Sons.
13. Ozsvath, D.L. (2009). Fluoride and environmental health: A review. *Rev. Environ. Sci. Biotechnol.*, 8, 59–79.
14. WHO (1984). Fluorine and fluorides. Environmental Health Criteria, 36. Geneva, Switzerland: World Health Organization.
15. Jacks, G., Bhattacharya, P., Chaudhary, V. and Singh, K.P. (2005). Controls on the genesis of some high fluoride groundwaters in India. *Appl. Geochem.*, 20, 221–228.
16. Umar, R. and Sami Ahmad, M. (2000). Groundwater quality in parts of Central Ganga Basin, India. *Environ. Geol.*, 39, 673–678.
17. Datta, K.K. (2000). Reclaiming salt-effected land through drainage in Haryana, India: A financial analysis. *Agric. Water Manag.*, 46, 55–71.
18. Ramamohana Rao, N.V., Suryaprakasa Rao, K. and Schuiling, R.D. (1993). Fluorine distribution in waters of Nalgonda District, Andhra Pradesh, India. *Environ. Geol.*, 21, 84–89.
19. Duraiswami, R.A. and Patankar, U. (2011). Occurrence of fluoride in the drinking water sources from gad river basin, Maharashtra. *J. Geo. Soc. India*, 77, 167–174.
20. ATSDR (2003). Report on Toxicological Profile For Fluorides, Hydrogen Fluoride and Fluorine. U.S. Department of Health and Human Services, Public Health Service Agency for Toxic Substances and Disease Registry.
21. WHO (2002). Fluorides, Environmental Health Criteria Number, WHO Monograph No. 227. Geneva, Switzerland: World Health Organization.
22. MRC (2002). Working Group Report: Water Fluoridation and Health, MRC: 47. London: Medical Research Council, 2002. http://www.mrc.ac.uk/pdf-publications-water_fluoridation_report.pdf.

23. Fomon, S.J. and Ekstrand (1999). Fluoride intake by infants. *J. Public Health Dent.*, 59, 229–234.
24. Oliveby, A., Twetman, S. and Ekstrand, J. (1990). Diurnal fluoride concentration in whole saliva in children living in a high- and a low-fluoride area. *Caries Res.*, 24, 44–47.
25. Li, Y., Liang, C.K., Slemenda, C.W., Ji, R., Sun, S., Cao, J., Emsley, C.L., Ma, F., Wu, Y., Ying, P., Zhang, Y., Gao, S., Zhang, W., Katz, B.P., Niu, S., Cao, S. and Johnston, C.C. Jr. (2001). Effect of long term exposure to fluoride in drinking water on risks of bone fractures. *J. Bone Miner. Res.*, 16, 932–939.
26. Saxena, V.K. and Ahmed, S. (2001). Dissolution of fluoride in ground water: A water-rock interaction study. *Environ. Geol.*, 40, 1084–1087.
27. Handa, B.K. (1975). Geochemistry and genesis of fluoride-containing ground waters in India. *Groundwater*, 13(3).
28. Saxena, V.K. and Shakeel, A. (2003). Inferring the chemical parameters for the dissolution of fluoride in ground water. *Environ. Geol.*, 43, 731–736.
29. Subba, R.N. and John, D.D. (2003). Fluoride incidence in ground water in an area of Peninsular India. *Environ. Geol.*, 45, 243–251.
30. Butler, J.N. (1964). *Ionic Equilibrium—A Mathematical Approach*. Reading, MA: Addison-Wesley Publishing Co., Inc.
31. Krishnamachari, K.A.V.R. (1976). Further observations on the syndrome of endemic genu valgum of South India. *Ind. J. Med. Res.*, 64, 284–291.
32. Singh, R.B. (2000). Environmental consequences of agricultural development: A case study from the green revolution state Haryana, India. *Agric. Ecosys. Environ.* 82, 97–103.
33. Apambire, W.B., Boyle, D.R. and Michel, F.A. (1997). Geochemistry, genesis, and health implications of fluoriferous ground waters in the upper regions of Ghana. *Environ. Geol.*, 33, 13–24.
34. He, X., Ma, T., Wang, Y., Shan, H. and Deng, Y. (2013). Hydrogeochemistry of high fluoride ground water in shallow aquifers, Hangjinhouqi, Hetao Plain. *J Geochem. Explor.* http://dx.doi.org/10.1016/j.gexplo.2012.11.010.
35. Rafique, T., Naseem, S., Usmani, T.H., Bashir, E., Khan, F.A. and Bhanger, M.I. (2009). Geochemical factors controlling the occurrence of high fluoride groundwater in the Nagar Parkar area, Sindh, Pakistan. *J. Hazard. Mater.*, 171, 424–430.
36. Mamatha, P. and Rao, S.M. (2010). Geochemistry of fluoride rich groundwater in Kolar and Tumkur Districts of Karnataka. *Environ. Earth Sci.*, 61, 131–142.
37. MHC (2010). *Ministry of Health of China, China Health Statistical Yearbook 2010*. Beijing: Peking Union Medical College Press.
38. Guo, Q.H., Wang, Y.X., Ma, T. and Ma, R. (2007). Geochemical processes controlling the elevated fluoride concentrations in groundwaters of the Taiyuan basin, Northern China. *J. Geochem. Explor.*, 93, 1–12.

2

Scenario of Fluoride Pollution

2.1 Introduction

Considerable bodies of scientific literature suggest that fluoride pollution has been spreading its tentacles into many regions of the world. As a result, more and more human habitations are forced to consume fluoride-rich groundwater. The situations in China and India, the most populous countries of the world, are the worst. In many regions of the world, especially those in the developing countries, the issue has acquired the dimensions of a socioeconomic problem rather than being one pertaining to mere water quality. Thus, it would be interesting to have a look into the progression of fluoride invasions over the boundaries of the world.

2.2 Global Scenario

Of late, fluorosis has become endemic in many parts of the world (Figure 2.1), as it has spread its tentacles into more than 40 nations, particularly into midlatitude regions. Groundwater with high fluoride occurs in large parts of Africa, China, the Middle East, and southern Asia, including India and Sri Lanka. The belt from Eritrea to Malawi along the East African Rift is the most popular fluoride belt on the earth. Other fluoride-rich zones span wide across China, Northern Thailand, India, Afghanistan, Iran, Iraq, and Turkey. The United States and Japan have similar fluoride belts, as shown in Figure 2.1.[1] India and China are the worst affected nations due to fluoride attacks (Figures 2.2 and 2.3). The global scenario of the affected countries along with a description on the reported intensity and severity of excess fluoride is presented in Table 2.1.

2.2.1 Asian and African Scenario

South Asia serves as home to nearly 25% of the world population, including 43% of the world's poor. Unfortunately, one-fourth of the populations in the developing world are living under acute water scarcity. The rural populations

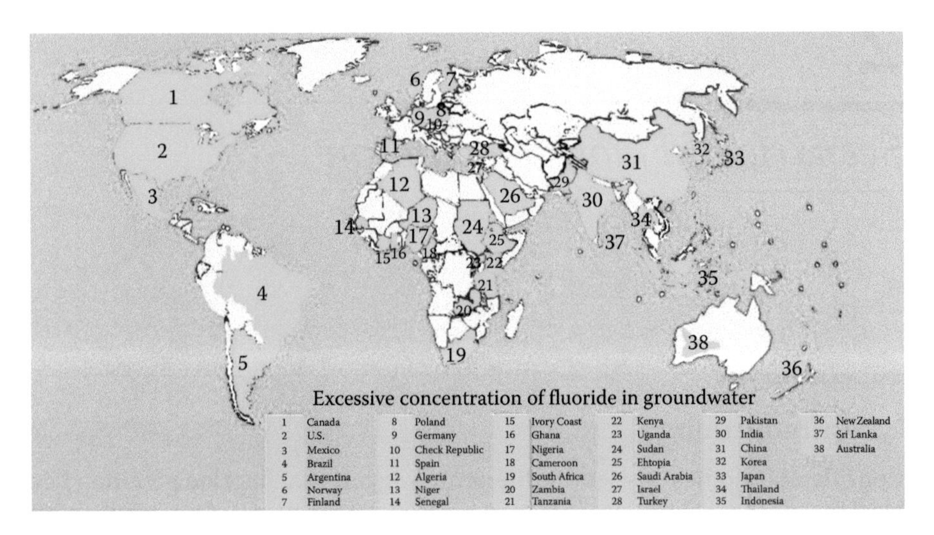

FIGURE 2.1
Global fluoride map. (Modified from WHO, *Fluoride and Arsenic in Drinking Water*, World Health Organization, Geneva, Switzerland, 2004, http://www.who.int/water_sanitation_health/en/map08b.jpg.)

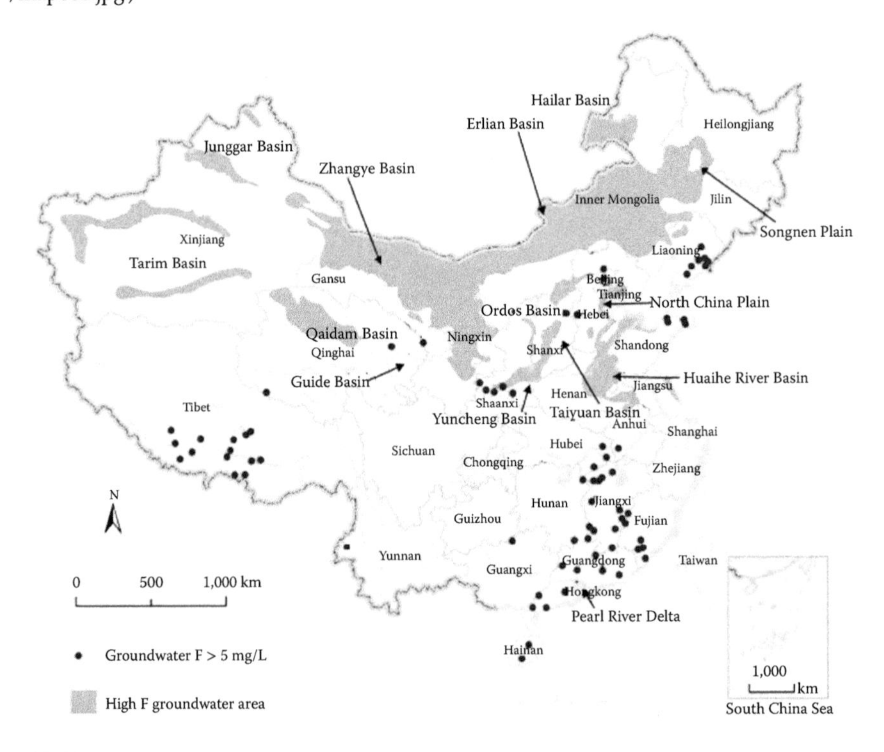

FIGURE 2.2
Fluoride map of China. (Modified from Wen, D., Zhang, F., Zhang, E., Wang, C., Han, S. and Zheng, Y., *J. Geochem. Explor.*, 135, 1–21, 2013.)

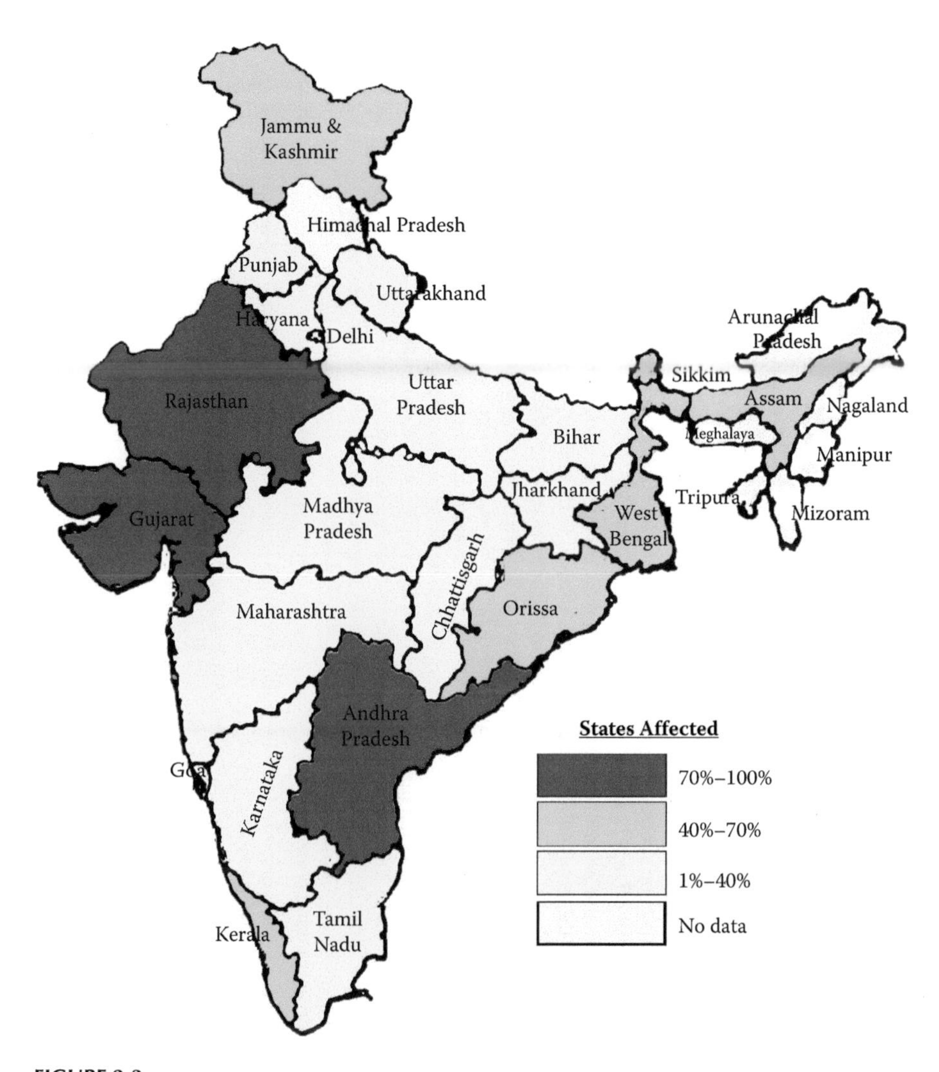

FIGURE 2.3
Fluoride map of India. (Modified from UNICEF, States of the Art Report on the Extent of Fluoride in Drinking Water and the Resulting Endemicity in India, Fluorosis and Rural Development Foundation for UNICEF, New Delhi, 1999.)

are very vulnerable to water-borne diseases due to contamination of drinking water; this ultimately leads to the outbreak of endemic diseases. The continued use of untreated water from shallow groundwater sources increases the disease burden of these rural habitations. As per the WHO assessment in 2004, two-thirds of people in Asia live without any access to drinking water. Further, the water quality issues due to the intrusion of geogenic pollutants such as fluoride into the scarce drinking water sources add to this misery. The intensity and gravity of the water quality issues that are persisting in these regions and associated habitations are so huge to quantify and tackle.[7, 69,70]

TABLE 2.1

Global Scenario of the Intensity and Severity of Excess Fluoride in Drinking Water

S. No	Affected Nations	Reported Fluoride Levels in Drinking Water and Associated Effects	References
1	India	In India, fluoride concentrations ranging from 0.5 to around 70 mg/L were reported. Andhra Pradesh, Rajasthan, and Gujarat are the severely affected states. Maximum fluoride concentrations reported include 69.7 mg/L in Rajasthan, 23 mg/L in Assam, 32 mg/L in New Delhi, and 48 mg/L in Haryana. Earlier, around 67 million people living in more than 20 states in India were estimated to be "at risk" of fluorosis. Crippling skeletal fluorosis was reported at a low fluoride concentration of 2.8 mg/L; dental fluorosis, at 0.5 mg/L; and skeletal fluorosis, at 0.7 mg/L.	Ayoob et al.,[5] Agarwal et al.,[6] Ayoob and Gupta,[7] Susheela,[8] Susheela and Bhatnagar[9]
2	China	In 1990, it was reported that around 300 million people in China were exposed to fluoride-rich waters and associated issues. Out of these, 40 million people had been afflicted with dental fluorosis and 3 million had been afflicted with skeletal fluorosis. However, in 1995, it was reported that one-tenth of the total Chinese population was exposed to endemic fluorosis. In Kuitan region of Zhuiger basin, concentrations to the level of 21.5 mg/L were reported. The data in 2004 suggested that more than 26 million people in China suffered from dental fluorosis and 1 million suffered from skeletal fluorosis. Fluorosis was extensively reported in China from Shanxi, Inner Mongolia, Shandong, Henan, and Xinjiang provinces due to high fluoride levels in drinking water. As per the endemic fluorosis control status of China in 2006, more than 1.34 million inhabitants suffer from skeletal fluorosis and 21.45 million suffer due to dental fluorosis. In 2010, there were 41.76 million fluorosis cases in 1325 different counties of China, out of which 58.2% were caused by chronic exposure to high levels of fluoride in drinking water.	Li and Cao,[10] UNICEF,[11] Wang et al.,[12] Wang and Huang,[13] WHO,[14,15] Chen,[16] MHC,[17] MHPRC[18]
3	Tanzania	Many regions of Tanzania are the worst affected with fluorosis-related issues. The range of reported fluoride concentrations varies from 8 to 12.7 mg/L. Severe cases of dental, skeletal, and crippling fluorosis were reported from Singida, Shinyanga, Mwanza, Kilimanjaro, Mara, and Arusha regions.	Mjengera and Mkongo[19]

(Continued)

TABLE 2.1 (*Continued*)

Global Scenario of the Intensity and Severity of Excess Fluoride in Drinking Water

S. No	Affected Nations	Reported Fluoride Levels in Drinking Water and Associated Effects	References
4	South Africa	The reported fluoride levels in South Africa vary from 0.05 to 13 mg/L. Concentrations as high as 30 mg/L were reported from Western Bushveld and Pilanesberg. Fluoride concentrations of 3, 0.48 and 0.19 mg/L were reported from Lee Gamka, Kuboes, and Sanddrif, respectively. The occurrences of dental fluorosis in children belonging to these regions were reported as 95%, 50% and 47%, respectively. In Western Bushveld regions, acute cases of skeletal fluorosis were reported. The morbidity rate of dental fluorosis in the Northwest province was similarly very high (97%).	Coetzee et al.,[20] Grobler and Dreyer,[21] Grobler et al.,[22] Mothusi[23]
5	Kenya	In Kenya, the reported fluoride concentration varies from 1 to 8.0 mg/L, with a fluorosis prevalence rate of 44%–77%. Issues and incidences of skeletal fluorosis were reported at fluoride levels of 18 mg/L. The highest fluoride levels were reported from the Rift Valley around Naivasha, Mount Kenya, and Nakuru and regions near the northern frontier, in addition to the peri-urban areas of Nairobi. Throughout Kenya, the local fluoride concentrations vary from 2 to 20 mg/L. Very high concentrations of 2800 and 1640 mg/L were reported from lakes of Nakuru and Elmentaita, respectively. In a sample study consisting of 1000 groundwater samples, more than 600 samples exceeded 1 mg/L, 200 samples crossed 5 mg/L, and more than 120 samples showed fluoride concentrations higher than 8 mg/L.	Kaimenyi,[24] Nair and Manji,[25] Nair et al.[26]
6	Ghana	In Ghana, 62% of school children were afflicted with dental fluorosis in the Bongo areas. The fluoride levels were found to be in the range of 0.11–4.6 mg/L. Recently, the presence of excess fluoride in groundwater (to the tune of 11.6 mg/L) is reported from the northern region of Ghana.	Apambire et al.,[27] Salifu et al.[28]
7	Sudan	In 1953, the reported fluoride concentrations in Abu Deleig and Jebel Gaili were in the range of 0.65–3.2 mg/L. Incidentally, the dental fluorosis in Abu Deleig was higher than 60%. In 1995, a high prevalence of (91%) dental fluorosis was observed among those children drinking water with 0.25 mg/L of fluoride.	Ibrahim et al.,[29] Smith and Smith[30]

(*Continued*)

TABLE 2.1 (*Continued*)

Global Scenario of the Intensity and Severity of Excess Fluoride in Drinking Water

S. No	Affected Nations	Reported Fluoride Levels in Drinking Water and Associated Effects	References
8	United States	The average fluoride concentrations reported in Illinois were 1.06 and 4.07 mg/L, and those in Texas were 0.3 and 4.3 mg/L. In the hot springs and geysers of the National Park at Yellowstone, fluoride levels of 25 to 50 mg/L were reported. The range of fluoride concentration in Lakeland at Southern California was 3.6–5.3 mg/L. A range of 5.0–15 mg/L of fluoride was reported in the deep aquifers of Western United States.	Cohen and Conrad,[31] Driscoll et al.,[32] Neuhold and Sigler,[33] Reardon and Wang,[34] Segreto et al.[35]
9	Mexico	In Mexico, around 6% of the population (around 5 million people) is affected by high concentrations of fluoride in groundwater. The maximum concentrations reported were in Abasolo in Guanajuato state (8 mg/L) and in Hermosillo in Sonara state (7.8 mg/L). Average fluoride levels of 0.9–4.5 mg/L were observed in rural locations and of 1.5–2.8 mg/L were observed in urban locations.	Díaz-Barriga et al.,[36] UNICEF[37]
10	Ethiopia	A high occurrence of dental fluorosis was reported in Ethiopian Rift Valley, where the concentrations of fluoride range from 1.5 to 177 mg/L. The Wonji-Shoa sugar estates in Ethiopian Rift Valley recorded the highest occurrence of skeletal and crippling skeletal fluorosis. In the Main Ethiopian Rift (MER) Valley (a part of the East African Rift), among 10 million people, more than 8 million are exposed to an elevated concentration of fluoride. It is estimated that about 1.2 million inhabitants drink groundwater with fluoride contents that exceed international guideline values.	Haimanot et al.,[38] Kloos et al.,[39] Rango et al.[40]
11	Canada	High fluoride concentrations were reported from Alberta (4.3 mg/L), Saskatchewan (2.8 mg/L) and Quebec (2.5 mg/L) in Canada. The range of fluoride concentrations observed in Rigolet and Labrador was 0.1–3.8 mg/L. Issues of dental fluorosis were also reported from Rigolet.	Health Canada, Priority Substances List Assessment Report on Inorganic Fluorides,[41] Ismail and Messer,[42] WHO[43]

(*Continued*)

TABLE 2.1 (*Continued*)

Global Scenario of the Intensity and Severity of Excess Fluoride in Drinking Water

S. No	Affected Nations	Reported Fluoride Levels in Drinking Water and Associated Effects	References
12	Poland, Finland, Czech Republic, Brazil, Indonesia, Israel, Turkey, Cameron, Zambia, and Europe	Fluoride concentrations higher than 3 mg/L were reported from Czech Republic, Finland, and Poland. In Brazil, higher concentrations of fluoride were reported from Paraiba state (0.1–2.3 mg/L) in the northeast region and from Ceara state (2–3 mg/L). In Indonesia, fluoride levels of 0.1–4.2 mg/L were reported in the well waters of the north-eastern part of Java in the Asembagus coastal plain. In Israel, natural fluoride concentrations up to 3 mg/L were reported from the Negev desert regions. In Turkey, high fluoride concentrations were reported from the Middle and Eastern parts. Denizli-Saraykoy and Caldiran plains exhibited a maximum concentration of 13.7 mg/L. Recently, high fluoride-induced dental and skeletal fluorosis have been observed in Cameroon, Zambia, and Europe.	Azbar and Türkman,[44] Cortes et al.,[45] Czarnowski et al.,[46] Heikens et al.,[47] Milgalter et al.,[48] WHO,[14,43] Oruc,[49] Fantong et al.,[50] Fordyce et al.[51] Shitumbanuma et al.[52]
13	Ivory coast, Senegal, North Algeria, Uganda, and Argentina	The occurrence of fluorosis was reported from all these regions. High prevalence of severe dental fluorosis among children (30%–60%) was reported in Senegal, from Guinguineo and Darou Rahmane Fall regions with fluoride concentrations of 4.6 and 7.4 mg/L. In Argentina, the south-east subhumid pampa regions showed fluoride levels of 0.9–18.2 mg/L with an average value of 3.8 mg/L. In Western Uganda, issues of dental fluorosis were reported from the Rift Valley area with fluoride levels of 0.5–2.5 mg/L.	Brouwer et al.,[53] Paoloni et al.,[54] Rwenyonyi et al.,[55] WHO[1]

(*Continued*)

TABLE 2.1 (*Continued*)

Global Scenario of the Intensity and Severity of Excess Fluoride in Drinking Water

S. No	Affected Nations	Reported Fluoride Levels in Drinking Water and Associated Effects	References
14	Norway, New Zealand, Germany, Spain, Niger, Nigeria, Pakistan, and Iran	The occurrence of fluorosis has been reported from all these regions. Issues of severe dental fluorosis were reported in the county of Hordaland, Norway, where fluoride levels reported in groundwater ranged from 0.02 to 9.48 mg/L. In the Muenster regions of Germany, fluoride concentrations of 8.8 mg/L were reported. The range of fluoride concentration in the Tenerife areas of Spain varies from 2.50 to 4.59 mg/L. The prevalence of skeletal fluorosis was reported from Tibiri in Niger; this was evident among boys who were exposed to fluoride concentrations of 4.7–6.6 mg/L. A fluoride exposure level of 0.5–3.96 mg/L in the Langtang town area of Nigeria exhibited 26.1% occurrence of dental fluorosis. In Pakistan, fluoride concentrations of 8–13.52 mg/L were observed in and around the spring and stream sources of Naranji. High fluoride levels (8.85 mg/L) in drinking water and high prevalence of dental fluorosis are reported from Iran.	WHO,[1] Bardsen et al.,[56] Hardisson et al.,[57] Queste et al.,[58] Shah and Danishwar,[59] Wongdem et al.,[60] Poureslami et al.,[61] Fekri and Kasmaei[62]
15	Saudi Arabia, Eritrea, Sri Lanka, Thailand, Japan, and Korea	Elevated fluoride levels were reported from Saudi Arabia in Mecca (2.5 mg/L) and Hail regions (2.8 mg/L) along with incidences of fluorosis. In Eritrea, elevated fluoride concentrations of 2.02–3.73 mg/L were reported in and around Keren areas. In Sri Lanka, very high concentrations up to 10 mg/L were reported in the North Central Province. In Thailand, at least 1% of the natural drinking water sources are laced with fluoride concentrations that are more than 2 mg/L, with higher values exceeding 10 mg/L. In Japan, people exposed to 1.4 mg/L fluoride concentrations are afflicted with 15.4% prevalence of dental fluorosis. One-fourth of the total wells in the south-eastern part of Korea have groundwater with fluoride concentrations greater than 5 mg/L.	Akpata et al.,[63] Al-Khateeb et al.,[64] Dissanayake,[65] Kim and Jeong,[66] Srikanth et al.,[67] Tsutsui et al.[68]

Source: Modified from Ayoob, S., Gupta, A. K. and Bhat, V.T., Critical Rev. *Environ. Sci. Technol.*, 38, 401–470, 2008.

High fluoride concentrations in drinking water and associated fluorosis issues were reported from African countries such as Tanzania, South Africa, Kenya, Ghana, and Sudan (Table 2.1). Tanzania is one of the countries in the world that is severely affected by fluorosis. The crippling skeletal fluorosis keeps people, especially children, immobile. Severe issues related to crippling skeletal and dental fluorosis were reported from regions such as Shinyanga, Singida, Mara, Mwanza, Kilimanjaro, and Arusha (Figure 2.4); however, Dodoma, Kigoma, Tabora, and Tanga are moderately affected. Reports dealing with excess fluoride in drinking water and associated health problems in South Africa made their appearance from 1935 onward (Figure 2.5). North-West provinces, Limpopo, Northern Cape, and major portions of Karoo are badly affected with issues of excess fluoride. Areas belonging to Western Bushveld and Pilanesberg reported fluoride concentrations higher than 1 mg/L. Even very high concentrations (around 30 mg/L) were reported in alkaline waters with a pH higher than 9. High fluoride concentrations exist in groundwater of North-West and Kwa-Zulu-Natal provinces, Northern Cape and Limpopo. In the North-West Province, the morbidity of dental fluorosis was 97%. Kenya in East Africa is bordered by Somalia, Ethiopia, Sudan, Uganda, and Tanzania. Fluorosis incidences of varying degrees were observed in Kenya, with a general prevalence rate up to 77%. Issues of skeletal fluorosis were identified among people drinking borehole water with high fluoride concentrations of 18 mg/L. In Ghana, more than 60% of the total populations of school children

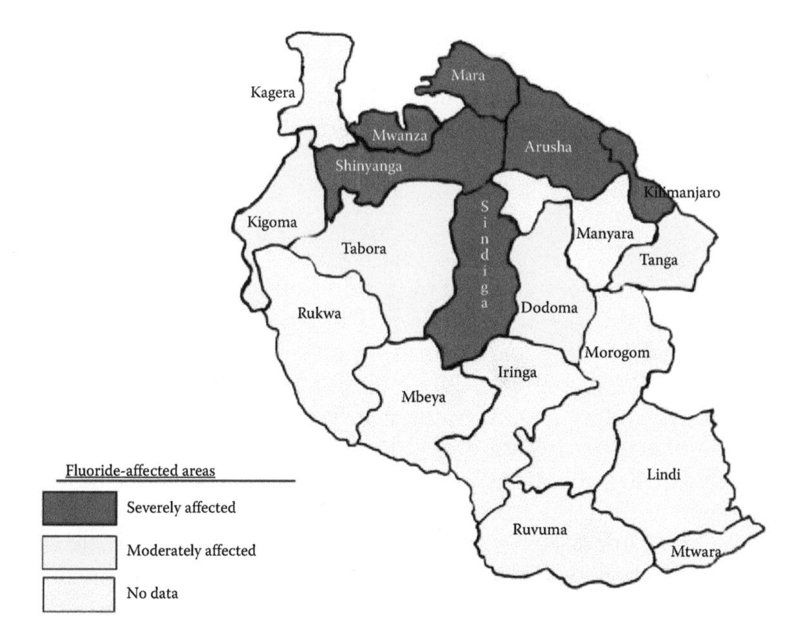

FIGURE 2.4
Fluorotic map of Tanzania. (Modified from Mjengera, H. and Mkongo, G., *Phys. Chem. Earth*, 28, 1097–1104, 2003.)

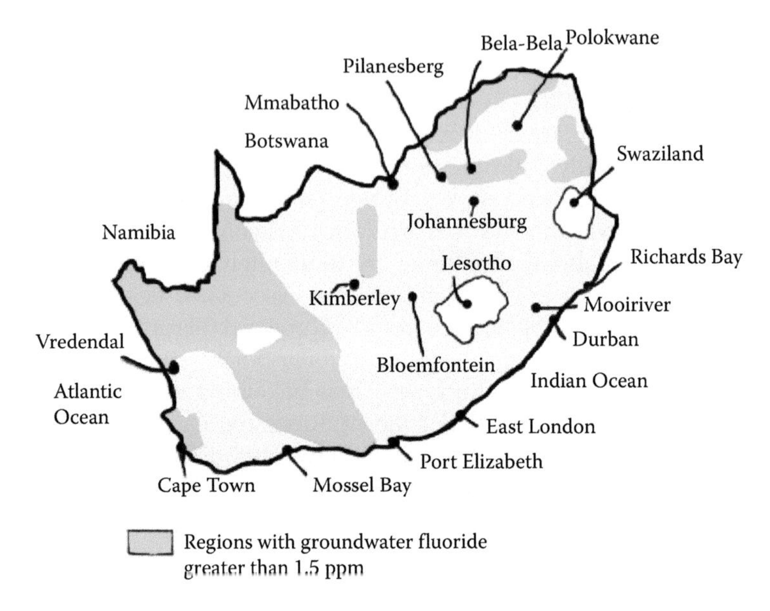

FIGURE 2.5
Fluorosis map of South Africa with groundwater fluoride concentration greater than 1.5 mg/L. (Modified from McCaffrey, L. P. and Willis, J. P., Distribution of Fluoride-Rich Groundwater in the Eastern and Mogwase Regions of the Northern and North-West Provinces, W.R.C. Report. No. 526/1/01, 2001.)

in the Bongo area reported to have dental fluorosis. In Bolgatanga and Bongo districts, elevated levels in natural groundwater were reported. In Sudan, 91% prevalence of dental fluorosis was observed among children consuming water having 0.25 mg/L of fluoride.[7]

2.2.2 Indian Scenario

India is the seventh largest, second most populous country in the world and is home to 16% of the global population. Though home to a sixth of humanity, India has just 4% of global water resources. The expected population is 1330 million in 2020. Though the bulk majority of the drinking water requirements are addressed by groundwater, nearly one-half of Indian villages are facing issues of acute drinking water shortage. Thus, providing reliable and potable water to all habitations is still a distant dream. Of late, the advancements made in India in the frontier areas of irrigation and food security were believed to be at the expense of groundwater. As per an estimate in 2004, around 3.7 billion bore wells were constructed for irrigation. The uncontrolled mining of groundwater through these bore wells contributed to an imbalance in our natural ecosystem. This imbalance is believed to be one of the reasons for the increased scarcity and pollution of our groundwater sources. Thus, such unfettered pumping from groundwater sources might have resulted in the decline of the natural water table, which might have triggered the entry of geogenic pollutants such as fluoride into groundwater aquifers. Further, geological processes governed by different hydrological and geochemical settings might have accelerated the entry of fluoride into groundwater.

In the 1930s, though fluorosis was reported from only 4 states of India, as per latest reports, more than 20 states are affected. Rajasthan, Andhra Pradesh, and Gujarat are the worst affected states (Figure 2.3). Higher fluoride concentrations of 44 mg/L and 23 mg/L, respectively, are reported from Rajasthan and Assam (Table 2.1). A natural maximum fluoride concentration of 32 mg/L was reported from the mega city of Delhi.[8] A fluoride concentration of 21.0 mg/L is reported in the groundwater at Kurmapalli watershed in Nalgonda district of Andhra Pradesh (the worst affected state of India); this is one of the highest concentrations in the groundwater of granite terrain in the country.[72] The granitic rocks in Nalgonda were identified as the chief source of excess fluoride in the groundwater that ranges from around 300 to 3200 mg/L. The granitic rocks of Nalgonda appear to have the highest fluoride content in the world with a mean fluoride concentration of 1440 mg/L.[73]

2.3 Summary

- The fluoride belt from Eritrea to Malawi along the East African Rift is the most popular fluoride belt on the earth. Other fluoride-rich zones span wide across China, Northern Thailand, India, Afghanistan, Iran, Iraq, and Turkey.
- More than 65% of people in Asia have no access to safe drinking water. Due to the intrusion of fluoride into drinking water sources, water quality issues could turn the Asian people's lives miserable.
- In addition to China and India, high fluoride concentrations in drinking water and associated fluorosis issues were reported from African countries such as Tanzania, South Africa, Kenya, Ghana, and Sudan. Tanzania is one of the countries in the world that is severely affected by fluorosis. The most recent literature suggests that people from more than 40 nations across the world are under the "risk" of fluorosis.
- Geological processes facilitating the weathering of rocks and associated fluoride-bearing minerals under different hydrological and geochemical settings accelerate the entry of fluoride into groundwater.
- Although fluorosis was reported from only 4 states of India in the 1930s, as per the latest reports, more than 20 states are affected. Rajasthan, Andhra Pradesh, and Gujarat are the worst affected states. Higher fluoride concentrations of 44 and 23 mg/L, respectively, are reported from Rajasthan and Assam.

References

1. WHO (2005). World Sanitation and Heath, Geneva, Switzerland: World Health Organization. http://www.who.int/water_sanitation_health/diseases/fluorosis/en.
2. WHO (2004). Fluoride and Arsenic in Drinking Water. Geneva, Switzerland: World Health Organization. http://www.who.int/water_sanitation_health/en/map08b.jpg.
3. Wen, D., Zhang, F., Zhang, E., Wang, C.L., Han, S. and Zheng, Y. (2013). Arsenic, fluoride and iodine in groundwater of China. *J. Geochem. Explor.*, 135, 1–21.
4. UNICEF (1999). States of the Art Report on the Extent of Fluoride in Drinking Water and the Resulting Endemicity in India. New Delhi, India: Fluorosis and Rural Development Foundation for UNICEF.
5. Ayoob, S., Gupta, A.K. and Bhat, V.T. (2008). A conceptual overview on sustainable technologies for the defluoridation of drinking water. *Crit. Rev. Environ. Sci. Technol.*, 38, 401–470.
6. Agarwal, C.K., Gupta, K.S. and Gupta, B.A. (1999). Development of new low cost defluoridation technology. *Water Sci. Technol.*, 40, 167–173.
7. Ayoob, S. and Gupta, A.K. (2006). Fluoride in drinking water: A review on the status and stress effects. *Crit. Rev. Environ. Sci. Technol.*, 36, 433–487.
8. Susheela, A.K. (2003). *A Treatise on Fluorosis*, 2nd revised edn, pp. 13–14. New Delhi, India: Fluorosis Research and Rural Development Foundation.
9. Susheela, A.K. and Bhatnagar, M. (1999). Structural aberrations in fluorosed human teeth: Biochemical and scanning electron microscopic studies. *Curr. Sci.*, 77, 1677–1680.
10. Li, J. and Cao, S. (1994). Recent studies on endemic fluorosis in China. *Fluoride*, 27, 125–128.
11. UNICEF. Web site: UNICEF statement on fluoride in water, http://www.fluoride.org.uk/statements/000000unicef.htm.
12. Wang, G.Q., Huang, Y.Z., Xiao, B.Y., Qian, X.C., Yao, H., Hu , Y., Gu, Y.L., Zhang, C. and Liu, K.T. (1997). Toxicity from water containing arsenic and fluoride in Xinjiang. *Fluoride*, 30, 81–84.
13. Wang, L.F. and Huang, J.Z. (1995). Outline of control practice of endemic fluorosis in China. *Soc. Sci. Med.*, 41, 1191–1195.
14. WHO (2006). In: Fawell, J., Bailey, K., Chilton, J., Dahi, E., Fewtrell, L. and Magara, Y. (Eds.), *Fluoride in Drinking Water*, pp. 41–75. London, UK: IWA Publishing.
15. WHO (2004). WHO issues revised drinking water guidelines to help prevent water-related outbreaks and disease, press release WHO/67. Geneva, Switzerland: World Health Organization.
16. Chen, N., Zhang, Z., Feng, C., Li, M., Chen, R. and Sugiura, N. (2011). Investigations on the batch and fixed-bed column performance of fluoride adsorption by Kanuma mud. *Desalination*, 268, 76–82.
17. MHC (2010). *Ministry of Health of China, China Health Statistical Yearbook 2010*. Beijing: Peking Union Medical College Press.
18. MHPRC (2007). Ministry of Health of the People's Republic of China, Chinese Health Statistical Digest. http://www.moh.gov.cn/open/2007tjts/P50.htm.

19. Mjengera, H. and Mkongo, G. (2003). Appropriate deflouridation technology for use in flourotic areas in Tanzania. *Phys. Chem. Earth*, 28, 1097–1104.
20. Coetzee, P.P., Coetzee, L.L., Puka, R. and Mubenga, I.S. (2003). Characterisation of selected South African clays for defluoridation of natural waters. *Water SA*, 29, 331–338.
21. Grobler, S.R. and Dreyer, A.G. (1988). Variations in the fluoride levels of drinking water in South Africa, Implications for fluoride supplementation. *S. Afr. Med. J.*, 73, 217–219.
22. Grobler, S.R., Dreyer, A.G. and Blignaut, R.J. (2001). Drinking water in South Africa: Implications for fluoride supplementation. *J. S. Afr. Dent. Assoc.*, 56, 557–559.
23. Mothusi, B. (1998). Psychological effects of dental fluorosis, Fluoride and Fluorosis, The Status of South African Research, Pilanesberg National Park, North West Province, 7, 1995; as cited in Muller, W.J., Heath, R.G.M. and Villet, M.H. (1998). Finding the optimum: Fluoridation of potable water in South Africa. *Water SA*, 24, 1–27.
24. Kaimenyi, T.J. (2004). Oral health in Kenya. *Int. Dent. J.*, 54, 378–382.
25. Nair, K.R. and Manji, F. (1982). Endemic fluorosis in deciduous dentition—A study of 1276 children in typically high fluoride area (Kiambu) in Kenya. *Odonto-Stomatol. Trop.*, 4, 177–184.
26. Nair, K.R., Manji, F. and Gitonga, J.N. (1984). The occurrence and distribution of fluoride in groundwaters in Kenya. *East Afr. J. Med.*, 61, 503–512.
27. Apambire, W.B., Boyle, D.R. and Michel, F.A. (1997). Geochemistry, genesis, and health implications of fluoriferous ground waters in the upper regions of Ghana. *Environ. Geol.*, 33, 13–24.
28. Salifu, A., Petrusevski, B., Ghebremichael, K., Buamah, R. and Amy, G. (2012). Multivariatestatistical analysis for fluoride occurrence in groundwater in the Northern region of Ghana. *J. Contam. Hydrol.*, 140–141, 34–44.
29. Ibrahim, Y.E., Affan, A.A. and Bjorvatn, K. (1995). Prevalence of dental fluorosis in Sudanese children from two villages with respectively 0.25 mg/L and 2.56 mg/L F in the drinking water. *Int. J. Paediatr. Dent.*, 5, 223–229.
30. Smith, H.R. and Smith, L.C. (1937). Bone contact removes fluoride. *Water Works Eng.*, 90, 600.
31. Cohen, D. and Conrad, M.H. (1998). 65,000 GPD fluoride removal membrane system in Lakeland, California, USA. *Desalination*, 117, 19–35.
32. Driscoll, W.S., Horowitz, H.S., Meyers, R.J., Heifetz, S.B., Kingman, A. and Zimmerman, E.R. (1983). Prevalence of dental caries and dental fluorosis in areas with optimal and above-optimal water fluoride concentrations. *J. Am. Dental Assoc.*, 107, 42–47.
33. Neuhold, J.M. and Sigler, W.F. (1960). Effects of sodium fluoride on carp and rainbow trout. *Trans. Am. Fish. Soc.*, 89, 358–370.
34. Reardon, J.E. and Wang, Y. (2000). A limestone reactor for fluoride removal from wastewaters. *Environ. Sci. Technol.*, 34, 3247–3253.
35. Segreto, V.A., Collins, E.M., Camann, D. and Smith, C.T. (1984). A current study of mottled enamel in Texas. *J. Am. Dental Assoc.*, 108, 56–59.
36. Díaz-Barriga, F., Navarro-Quezada, A., Grijalva, M.I., Grimaldo, M., Loyola-Rodriguez, J.P. and Ortz, M.D. (1997). Endemic fluorosis in Mexico. *Fluoride*, 30, 233–239.
37. UNICEF. Web site: UNICEF statement on fluoride in water, http://www.fluoride.org.uk/statements/000000unicef.htm.

38. Haimanot, R.T., Fekadu, A. and Bushra, B. (1987). Endemic fluorosis in the Ethiopian Rift Valley. *Tropic. Geogr. Med.*, 39, 209–217.
39. Kloos, H., Tekle-Haimanot, R., Fluorosis, Kloos, H. and Zein, A.H. (1993). *The Ecology of Health and Disease in Ethiopia*, pp. 44–541. Boulder, CO: Westview Press.
40. Rango, T., Bianchini, G., Beccaluva, L. and Tassinari, R. (2010). Geochemistry and water quality assessment of central Main Ethiopian Rift natural waters with emphasis on source and occurrence of fluoride and arsenic, *J. Afr. Earth Sci.*, 57, 479–491.
41. Health Canada, Priority Substances List Assessment Report on Inorganic Fluorides. (1993). *Canadian Environmental Protection Act, Minister of Supply and Services Canada*, pp. 12–19. Ottawa, Canada: Canada Communication Group-Publishing. K1A 0S9 1993.
42. Ismail, A.I. and Messer, J.G. (1996). The risk of fluorosis in students exposed to a higher than optimal concentration of fluoride in well water. *J. Public Health Dent.*, 56, 22–27.
43. WHO (2002). *Fluorides. Environmental Health Criteria*. WHO monograph No. 227. Geneva: World Health Organization.
44. Azbar, N. and Türkman, A. (2000). Defluoridation in drinking water. *Water Sci. Technol.*, 42, 403–407.
45. Cortes, D.F., Ellwood, R.P., O'Mullane, D.M. and de Magalhaes Bastos, J.R. (1996). Drinking water fluoride levels, dental fluorosis and caries experience in Brazil. *J. Public Health Dentistry*, 56, 226–228.
46. Czarnowski, W., Wrzesniowska, K. and Krechniak, J. (1996). Fluoride in drinking water and human urine in Northern and Central Poland. *Sci. Total Environ.*, 191, 177–184.
47. Heikens, A., Sumarti, S., vanBergen, M., Widianarko, B., Fokkert, L., van Leeuwen, K. and Seinen, W. (2005). The impact of the hyperacid Ijen Crater Lake: Risks of excess fluoride to human health. *Sci. Total Environ.*, 346, 56–69.
48. Milgalter, N., Zadik, D. and Kelman, A.M. (1974). Fluorosis and dental caries in Yotvata area. *Israel Dental J.*, 23, 104–109.
49. Oruc, N. (2008). Occurrence and problems of high fluoride waters in Turkey: An overview. *Environ. Geochem. Health*, 30, 315–323.
50. Fantong, W.Y., Satake, H., Ayonghe, S.N., Suh, E.C., Adelana, S.M.A., Fantong, E.B.S., Banseka, H.S., Gwanfogbe, C.D., Woincham, L.N., Uehara, Y. and Zhang, J. (2009). Geochemical provenance and spatial distribution of fluoride in groundwater of Mayo Tsanaga River Basin, Far North Region, Cameroon: Implications for incidence of fluorosis and optimal consumption dose. *Environ. Geochem. Health*, 32, 147–163.
51. Fordyce, F.M., Vrana, K., Zhovinsky, E., Povoroznuk, V., Toth, G., Hope, B.C., Iljinsky, U. and Baker, J. (2007). A health risk assessment for fluoride in Central Europe. *Environ. Geochem. Health*, 29, 83–102.
52. Shitumbanuma, V., Tembo, F., Tembo, J.M., Chilala, S. and Van Ranst, E. (2007). Dental fluorosis associated with drinking water from hot springs in Choma district in southern province, Zambia. *Environ. Geochem. Health*, 29, 51–58.
53. Brouwer, I.D., Dirks, O.B., De-Bruin, A. and Hautvast, J.G. (1988). Unsuitability of World Health Organization guidelines for fluoride concentrations in drinking water in Senegal. *Lancet*, 30, 223–225.

54. Paoloni, J.D., Fiorentino, C.E. and Sequeira, M.E. (2003). Fluoride contamination of aquifers in the southeast sub humid pampa, Argentina. *Environ. Toxicol.*, 18, 317–320.
55. Rwenyonyi, C.M., Birkeland, J.M., Haugejorden, O. and Bjorvatn, K. (2000). Age as a determinant of severity of dental fluorosis in children residing in areas with 0.5 and 2.5 mg fluoride per liter in drinking water. *Clin. Oral Invest.*, 4, 157–161.
56. Bardsen, A., Klock, K.S. and Bjorvatn, K. (1999). Dental fluorosis among persons exposed to high- and low-fluoride drinking water in western Norway. *Commun. Dentistry Oral Epidemiol.*, 27, 259–267.
57. Hardisson, A., Rodriguez, M.I., Burgos, A., Flores, L.D., Gutierrez, R. and Varela, H. (2001). Fluoride levels in publically supplied and bottled drinking waters in the island of Tenerife, Spain. *Bull. Environ. Contamin. Toxicol.*, 67, 163–170.
58. Queste, A., Lacombe, M., Hellmeier, W., Hillermann, F., Bortulussi, B., Kaup, M., Ott, K. and Mathys, W. (2001). High concentrations of fluoride and boron in drinking water wells in the Muenster region—Results of a preliminary investigation. *Int. J. Environ. Health*, 203, 221–224.
59. Shah, M.T. and Danishwar, S. (2003). Potential fluoride contamination in the drinking water of Naranji area, Northwest Frontier Province, Pakistan. *Environ. Geochem. Health*, 25, 475–481.
60. Wongdem, J.G., Aderinokun, G.A., Sridhar, M.K. and Selkur, S. (2000). Prevalence and distribution pattern of enamel fluorosis in Langtang town, Nigeria. *Afr. J. Med. Medical Sci.*, 29, 243–246.
61. Poureslami, H.R., Khazaeli, P. and Nooric, G.R. (2008). Fluoride in food and water consumed in koohbanan. *Fluoride*, 41, 216–219.
62. Fekri, M. and Kasmaei, L.S. (2011). Fluoride pollution in soils and waters of Koohbanan region, southeastern Iran. *Arab. J. Geosci.*, 6, 157–161.
63. Akpata, E.S., Fakiha, Z. and Khan, N. (1997). Dental fluorosis in 12–15-year-old rural children exposed to fluorides from well drinking water in the Hail region of Saudi Arabia. *Comm. Dentistry Oral Epidemiol.*, 25, 324–327.
64. Al-Khateeb, T.L., Al-Marasafi, A.I. and O'Mullane, D.M. (1991). Caries prevalence and treatment need amongst children in an Arabian community. *Commun. Dentistry Oral Epidemiol.*, 19, 277–280.
65. Dissanayake, C.B. (1996). Water quality and dental health in the Dry Zone of Sri Lanka. *Environ. Geochem. Health*, 113, 131–140.
66. Kim, K. and Jeong, Y.G. (2005). Factors influencing natural occurrence of fluoride-rich groundwaters: A case study in the southeastern part of the Korean Peninsula. *Chemosphere*, 58, 1399–1408.
67. Srikanth, R., Viswanatham, K.S., Kahsai, F., Fisahatsion, A. and Asmellash, M. (2002). Fluoride in groundwater in selected villages in Eritrea (North East Africa). *Environ. Monit. Assess.*, 75, 169–177.
68. Tsutsui, A., Yagi, M. and Horowitz, A.M. (2000). The prevalence of dental caries and fluorosis in Japanese communities with up to 1.4 ppm of naturally occurring fluoride. *J. Public Health Dentistry*, 60, 147–153.
69. WHO (2004). Ground Water and Public Health, Chapter 1. Geneva, Switzerland: World Health Organization. www.who.int/entity/water sanitation health /resources quality/en/groundwater1.pdf.
70. WHO (2004). FACTS Water, Sanitation and Hygiene Links to Health, Facts and Figures. Geneva, Switzerland: World Health Organization.

71. McCaffrey, L.P. and Willis, J.P. (2001). Distribution of Fluoride-Rich Groundwater in the Eastern and Mogwase Regions of the Northern and North-West Provinces. W.R.C. Report. No. 526/1/01.
72. Mondal, N.C., Prasad, R.K., Saxena, V.K., Singh, Y. and Singh, V.S. (2009). Appraisal of highly fluoride zones in groundwater of Kurmapalli watershed, Nalgonda district, Andhra Pradesh (India). *Environ. Earth Sci.*, 59, 63–73.
73. Ramamohana Rao, N.V., Rao, N., Rao, S.P. and Schuiling, H.D. (1993). Fluorine distribution in waters of Nalgonda District, Andhra Pradesh, India. *Environ. Geology*, 21, 84–89.

3

Dental Fluorosis

3.1 Introduction

Many studies suggest that fluoride may be an essential element for both animals and humans. However, it is true that the essentiality of fluoride for humans has not been demonstrated indisputably. Further, data on the minimum nutritional requirement are also inadequate. Incidentally, many epidemiological studies have clearly demonstrated possible adverse effects and health issues that arise due to the continuous ingestion of fluoride that is derived mainly through drinking water. These studies clearly show that fluoride primarily produces effects on skeletal tissues, especially bones and teeth. However, low concentrations of fluoride provide protection against dental caries, especially in children. According to the World Oral Health Report 2003, for a considerable percentage of people, especially children, in most of the industrialized countries, dental decay (dental caries) still remains a major public health issue. The changing living conditions and dietary habits are expected to be reasons for increased incidences of dental decay. Although considerable advancements have been made in preserving and improving global oral health issues, many of such issues related to the poor, marginalized, and disadvantaged groups still persist. Scientific research on the oral health issues related to fluoride started more than a century ago and focused on establishing a link between fluoride, dental caries, and fluorosis. Studies suggest that fluoride toothpastes and mouth rinses can significantly reduce the occurrence and prevalence of dental decay.[1]

3.2 Dental Effects of Fluoride

The presence of fluoride in water has dual significance on human health, that is, the concentration of fluoride in the drinking water defines both its beneficial and harmful effects. Although a minimum level of fluoride in drinking water may reduce dental caries, as stated earlier, higher concentrations may initiate dental fluorosis in various proportions. Though water fluoridation at minimal levels may be beneficial, minimizing the adverse fluorotic effects on teeth at higher concentrations should be cautioned. Accordingly, the optimum concentration of fluoride in drinking water is generally limited to 1 mg/L, as it may ensure maximum dental protection without initiating any adverse health problems.

Fluoride was considered to improve the crystal lattice stability of enamel and render it less soluble to acid demineralization. Since fluoride is incorporated into enamel as partially fluoridated hydroxyapatite, it is considered best when it is ingested. However, an increasing body of evidence suggests that a substantial part of the cariostatic activity of fluoride is due to its effects on erupted teeth. Further, the mechanism of action is mainly centered on the presence of fluoride in the fluid phase of dental plaque and saliva. The fluoride available in saliva and dental plaque reduces the demineralization of teeth and enhances the remineralization, mainly through an interaction with the surface of the enamel. Fluoridated toothpastes, mouth rinses, and topically applied dental treatments such as varnishes, gels, and solutions are also used in addition to water fluoridation. In addition, fluoridated milk, fluoridated salt, and other fluoride supplements are also used in many countries. Of late, considering the average annual maximum daily air temperature of the region, the optimum recommended range of fluoride in drinking water is from 0.5 to 1.2 mg/L.[2] The population using fluoride toothpastes (containing 10 g of fluoride per kg) for preventing dental caries may be twice as those consuming groundwater with excess fluoride. It is suggested that in many developed countries, use of these products is considered helpful for the gradual decline in the prevalence of dental decay. However, many studies suggest that fluoride supplements have only a limited role in enhancing dental health. It is pointed out in some studies that applications of fluoridated mouth rinses have become popular among school children and kids. However, their efficacy in preventing dental decay is dependent on the frequency of usage, level of compliance, and exposure to other fluoride sources. Proper caution is to be maintained in recommending the usage of mouth rinses for children younger than 6 years, as they may swallow it during rinsing, thereby increasing the risk of dental fluorosis.[2] Thus, it would be appropriate to recommend a mouth rinse only for the elderly people with elevated risks of dental decay.

3.2.1 Dental Caries (Tooth Decay)

Dental caries (decay) is one of the most prevalent chronic childhood diseases worldwide. The disease develops in both the crowns and roots of teeth, and it grows into aggressive tooth decay. It is presumed that there exists a physiological stability between the oral microbial films and teeth minerals. The alterations in this stable equilibrium may contribute to the initiation of dental caries.[3] It could be viewed as an "infectious and multifactorial disease" that is characterized by demineralization of inorganic components of teeth and dissolution of organic substances of microbial etiology. Unhygienic oral cavities invite bacterial growth, resulting in caries, which further leads to acid production by fermentation. This may etch away enamel, leaving black spots or cavities on teeth. These microorganisms can damage the soft pulp tissues by infiltrating through the dentin.[3] Dental decay ultimately leads to a state of acute systemic infection through different phases such as devastating pain, bacterial contagion resulting in pulpal necrosis, reduced dental function, tooth extraction, and so on. *Streptococcus mutans* and lactobacilli in dental plaque are mainly the two types of specific bacteria that are responsible for inducing caries.

3.2.2 Prevention of Dental Caries by Fluoride

The mineral crystals of calcium and phosphate form enamel and dentin, which are entrenched in an organic protein–lipid medium. Fluoride is utilized by cariogenic bacteria for fermentation, resulting in the production of acids. The absorbed/ingested fluoride inside the bacterial cell can interfere and alter the enzymic activities of the bacterial cell. Thus, bacterial activities and the resulting acid production get disturbed so that demineralization of dental mineral gets reduced. Incidentally, the adsorbed fluoride attracts calcium ions in saliva. Thus, fluoride aids the calcium and phosphate ions and chemically produces a crystal surface that is more resistant against acid solubility than the natural tooth mineral, thereby enhancing remineralization of the teeth.[4–6]

3.2.3 Role of Fluoride in Dental Decay

The presence of fluoride in water has dual significance. Since fluoride has both positive and negative impacts on human health, the stressful effects of fluoride in drinking water constitute a subject of intense scientific deliberation. As a result, many research findings in this direction are either contradicting or inconclusive. The population consuming fluoridated drinking water increased from 210 million in 1994 to 355 million in 2005.[2,7] In addition, it was estimated that more than 50 million people are consuming water with a naturally fluoridated concentration of 1 mg/L.[7] A systematic review investigating the association between fluoride and dental caries suggested that fluoridation of drinking water supplies reduces the occurrence of dental decay. Furthermore, since fluoride was found to be highly valuable when continually present at lower concentrations in plaque fluid and saliva, the WHO recommended the use of fluoridated mouth rinses to reduce elevated risk of dental caries.[2]

The effectiveness of fluoride in preventing tooth decay is a topic of intense scientific debate. Dental caries in children is said to be a bacterial disorder.[3] Several factors, namely, nutrition, oral bacteria, oral hygiene, and educational and economic statuses of parents, are cited as reasons for poor dental health. Large temporal reductions in tooth decay could be attributed to dietary patterns and immune status of populations. It is suggested that dietary control of caries without the use of fluoride is possible, as even chewing cheese reduces tooth decay.[8] It is also recommended that fluoride need not be treated as an essential nutrient to humans, as its absence has not proved to result in any disease. Since the essentiality of fluoride in humans is not established unambiguously, data on the minimum nutritional requirement of fluoride are unvailable.[6,9] It was also pointed out that majority of teeth decay issues develop on fissures and pits of the teeth, areas where fluoride is proved ineffective. This further demonstrates the topical action of fluoride on the teeth surface. Thus, many studies reserve apprehensions in actually consuming or ingesting fluoride.[10]

3.3 Dental Fluorosis: History and Occurrence

Dental fluorosis may appear as a cosmetic effect that ranges in appearance from scarcely discernible to a marked staining or pitting of the teeth in severe forms.[11] However, it could be treated as an early sign of fluoride attack that is visible to the naked eye; it also induces an irreversible toxic effect on tooth-forming cells. Although it histologically represents a hypocalcification, clinically it ranges from barely visible white striations on the teeth to gross defects and staining of teeth enamel.[11] The first report on dental fluorosis was from Mexico in 1888, where a family from Durango was identified with "black teeth." In 1891, cases of dental fluorosis were reported among Italian migrants from Naples to the United States. From 1900 onward, there was a pouring in of the prevalence and issues of dental fluorosis from different parts of the world, including the United States.[12] Dental fluorosis was first related to drinking water in 1925, though it was shown to be "specifically caused by fluoride in drinking water much before." It was Dr. Frederick S. McKay who first reported the development of an unusual permanent stain or "mottled enamel" on the teeth surface. This initiated and channelized a lot of scientific research on the relationship between fluoride in drinking water and fluorosis. In 1930, fluorosis was rated and identified as an "occupational disease in human." Subsequently, in 1932, in Denmark, the prevalence of skeletal fluorosis in cryolite miners was reported. In the 1930s, in India, fluorosis was first detected among cattle by the farmers of Nellore district of Andhra Pradesh. In 1937, the first medical report to this effect was published in the *Indian Medical Gazette*. In 1978, it was suggested that "fluorosis might be one of the most widespread endemic health problems" associated with natural geochemistry.[6,13–15]

3.4 Development of Dental Fluorosis

The progression of dental fluorosis is believed to take place through multiple stages. Initially, teeth become opaque and chalky due to "subsurface hypomineralization." Subsequently, the teeth lose enamel, which initiates the development of grooves and pits. Enamel and dentin, the calcium-rich constituents of teeth, have much affinity for fluoride in the formation and development of teeth. During mineralization of teeth, fluoride essentially combines with calcium to form calcium fluorapatite crystals. Due to such fluoride accumulation, calcium gets lost from teeth. As a result of this loss of calcium, fluorosed teeth move from a "mild to severe" state. Since calcium ions are lost to fluoride, teeth become weaker. Further, the severely fluorosed enamel becomes more discolored, pitted, porous, and prone to wear and fracture, as the "well mineralized zone is very fragile to mechanical stress." Due to the reduction in mineral content, mutilation of enamel mineralization, and associated structural alterations coupled with the morphological abnormalities on the teeth surface, the fluorosed teeth readily and easily get fractured.[3,16]

Most of the recent research suggests that "dental fluorosis results from a fluoride-induced delay in the hydrolysis and removal of amelogenin matrix proteins during enamel maturation and subsequent effects on crystal growth."[6] The proteins secreted by ameloblasts are known as amelogenins. These amelogenins slow down the growth of enamel crystallites. Once amelogeninases remove the amelogenins from the enamel matrix, the crystallite growth is increased in the initial maturation phase of tooth development. This most crucial enamel maturation phase turns out to be the most sensitive time for a higher level of fluoride exposure.[6,17,18] Depending on the level of exposure and nutritional status of the child, Dean (1934) classified the fluorosis on a scale from 0 to 4 as follows:[19] "class 0, no fluorosis; class 1, very mild fluorosis (opaque white areas irregularly covering about 25% of the tooth surface); class 2, mild fluorosis (white areas covering about 50% of the tooth surface); class 3, moderate fluorosis (all surfaces affected, with some brown spots and marked wear on surfaces subject to attrition); and class 4, severe fluorosis (widespread brown stains and pitting)."[6] Figure 3.1 shows the classification of dental fluorosis symptoms into seven categories according to Dean's classification (normal, questionable, very mild, mild, moderate, moderately severe, and severe), and each of these seven categories was given a numerical weight between 0 and 4.[20] Mild dental fluorosis is making its presence felt through the manifestation of small white areas in the enamel. Teeth that are stained and pitted or mottled in appearance denote the severe form of dental fluorosis. The hypomineralization of the enamel is the most prominent feature in human fluorotic teeth. The excessive fluoride assimilation into the enamel may interfere with its maturation process, resulting in alterations in the rheologic structure of the enamel matrix

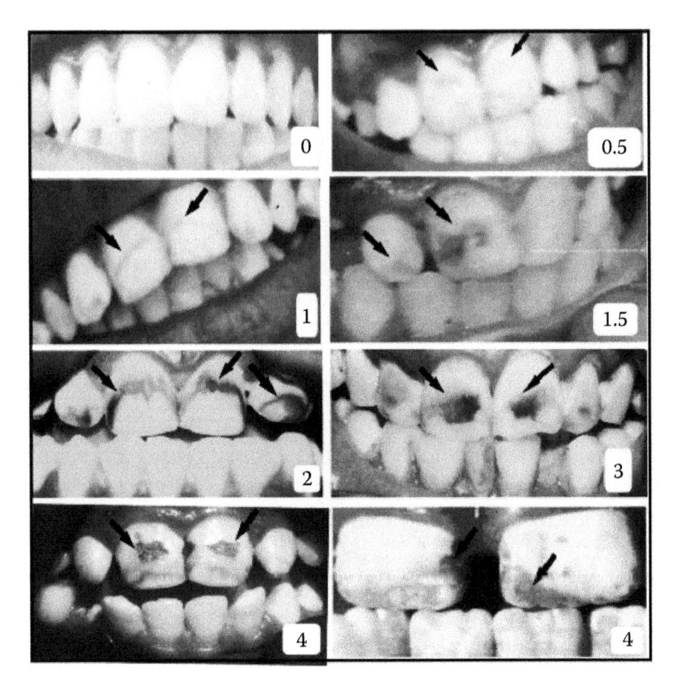

FIGURE 3.1
Dental fluorosis symptoms as per Dean's classification. (From Viswanathan, G., Jaswanth, A., Gopalakrishnan, S., Ilango, S. S. and Aditya, G., *Sci. Total Environ.*, 407, 5298–5307, 2009.)

and/or effects on cellular metabolic processes that are associated with normal enamel development.[21,22]

3.4.1 Physical Symptoms of Dental Fluorosis

As shown in Figure 3.1, the color of teeth may progress from white, yellow, and brown to black due to dental fluorosis. This "discoloration may be in spots, or appear as streaks invariably horizontal in orientation, as new layers of the matrix are added on horizontally" during the tooth development process.[3,6] Thus, the discoloration will usually appear in pairs based on developmental patterns and will not appear as a single isolated tooth. As against normal belief, the discoloration appears away from the gums, that is, on the enamel. As it may progress, in due course, the enamel will lose its brightness, luster, and shine. However, the discoloration due to other factors (e.g., dirty teeth, smoking, tobacco chewing, coffee or tea stains) may appear along the gums and only on the periphery of the teeth. Because "enamel lines are laid down in incremental lines during prenatal and postnatal periods," dental fluorosis may appear invariably as horizontal lines or bands (but never as vertical bands) on the teeth surface.[3,6]

FIGURE 3.2
A girl from Garhtipli village in Dhar district in Madhya Pradesh suffering from dental fluorosis. (From UNICEF, Photo essay, UNICEF/India/2006/Ruhani kaur, fluorosis-mitigating the scourge, http://www.unicef.org/india/1425.html.)

3.4.2 Issues of Dental Fluorosis

Teeth are important components of our facial skeleton, as they are vital in aesthetics, phonation, and speech. An attractive smile is the biggest human asset; however, a person with dental fluorosis gets deprived of this privilege. Dental fluorosis may impart psycho-sociological problems that arise due to distress, self-insolence, and loneliness. Children with dental fluorosis may be afraid to laugh, fearing embarrassment, and isolation, which may ruin their one and only childhood and their self-esteem. Such frustrations may eventually lead to deep psychological depressions. The fluoride endemic areas of the developing world agonizingly accommodate many human beings in this direction.[3,23,24] Figure 3.2 shows a girl from a rural Indian village who is suffering from dental fluorosis.

3.4.3 Prevalence of Dental Fluorosis

The correlation between the concentration of fluoride in water and the prevalence of dental fluorosis is well documented. As a consequence of the increased fluoride intake through multiple sources, the severity of dental fluorosis is increasing the world over. Only 50% of absorbed fluoride is retained in adults, whereas 80% is retained in children. Thus, children and adolescents become more susceptible to dental caries. Indian literature showed 100% prevalence of dental fluorosis at a fluoride level of 3.4–3.8 mg/L.[3,26] The "prevalence of dental fluorosis at a water fluoride level of 1 mg/L was estimated to be 48% in fluoridated areas and 15% in nonfluoridated areas. Limiting consideration to aesthetically important levels of severity, the prevalence of fluorosis is 12.5% in fluoridated areas and 6.3% in nonfluoridated areas. Increasing the water fluoride level from 0.4 to 1 mg/L, would mean that one additional person for every 22 people would have fluorosis of aesthetic concern, but with no risk."[6,27] Thus, it could be undoubtedly inferred that the benefit of reduction in dental decay due to fluoride is at the expense of the increased prevalence of dental fluorosis. A 100% prevalence of dental fluorosis was reported at a daily total fluoride intake of 2.78 mg/child/day. Further, at higher levels of daily total intake of fluoride, the prevalence of dental caries increased. It is important to monitor total fluoride exposure of children and excessive fluoride intake, especially during the years of tooth development.[28] The elevated fluoride content of the surrounding geological environment also imparts rich prevalence to dental fluorosis. Children in the age group of 3–14 years are more likely at the risk of fluorosis from consumption of vegetables and cereals grown in fluoridated areas.[29] This suggests the need for revising the maximum permissible limits of fluoride in drinking water prescribed by different regulatory bodies where sizable fluoride intake takes place through the food chain.

On examining more than 480,000 students covering 18 districts of Gujarat[6] (one of the three worst affected states in India), the percentage prevalence of dental fluorosis was found to vary from 2.6% to 33%. In Rajasthan (India), "maximum prevalence of dental fluorosis (77.1%) was observed among 17–22 year age group with severe dental fluorosis with black staining at 2.6 mg/L" fluoride.[6] The highest overall prevalence of dental fluorosis (77.2%) was observed at 3.2 mg/L. The study in Haryana (India) also demonstrates the correlation between dental fluorosis and dental caries in certain fluoride endemic locations. The "prevalence rate of dental fluorosis increases from 13% to 77% with increase in fluoride level from 0.64 mg/L to a range of 1.89–3.83 mg/L."[6] Incidentally, there is a steady increase in dental caries from 65% to 98%. This experimental evidence substantiates the poor correlation of dental caries with groundwater fluoride concentration levels. However, in Kolar district of Karnataka, India, genu valgum was prevalent among children with and without dental fluorosis.[30] These observations reveal that the prevalence and occurrences of fluorosis widely vary in populations at different regions, though they consume waters with almost the same fluoride concentrations. Other factors such as climate, individual biological responses, nutritional status, individual vulnerability, extent of fluoride exposure, and presence of other dissolved salts in the groundwater also have significant roles.[3,27,31]

3.5 Summary

- Literature suggests that fluoride may be an essential element for both animals and humans. However, for humans, the essentiality of fluoride has not yet been demonstrated indisputably. Thus, data indicating the minimum nutritional requirement of fluoride are unavailable.
- Many epidemiological studies have demonstrated possible adverse effects and health issues arising due to long-term ingestion of fluoride through drinking water. These studies clearly show that fluoride primarily produces effects on skeletal tissues, especially on bones and teeth.
- Considerable literature suggests that the usage of different water fluoridation techniques, applications of mouth rinses, and fluoride toothpastes significantly reduce the prevalence of dental caries.
- The acceptable and recommended permissible level of fluoride in drinking water for preventing dental caries ranges from 0.5 to 1.2 mg/L. This range of limit is fixed based on the annual average maximum daily air temperature of a region. In general, regulatory agencies suggest a general acceptable maximum fluoride concentration of 1 mg/L in drinking water. This is to ensure the maximum level of protection of dental caries and to avoid the prevalence of dental fluorosis.
- The ecological imbalance between physiological equilibrium of tooth minerals and oral microbial biofilms is a causative factor for dental caries.
- The effectiveness of fluoride in preventing tooth decay is a topic of intense scientific debate, as many studies reserve apprehensions in actually consuming or ingesting fluoride.
- Dental fluorosis could be treated as an early sign of fluoride attack that is visible to the naked eye. It induces an irreversible toxic effect on tooth-forming cells.
- It could be inferred that fluorosis may be regarded as one of the most pervasive endemic health problems generated and coupled with natural geochemistry.
- Most of the recent research suggests that "dental fluorosis results from a fluoride-induced delay in the hydrolysis and removal of

amelogenin matrix proteins during enamel maturation and subsequent effects on crystal growth."

- Dental fluorosis may impart psycho-sociological problems arising due to distress, self-insolence, and loneliness. Children with dental fluorosis may be afraid to laugh, fearing embarrassment, and isolation, which may ruin their one and only childhood and their self-esteem.
- Indian literature demonstrates a 100% prevalence of dental fluorosis at a fluoride level of 3.4–3.8 mg/L.
- Considerable literature suggests the need for revising the maximum permissible limits of fluoride in drinking water prescribed by different regulatory bodies where sizable fluoride intake takes place through the food chain.

References

1. Petersen, P.E. and Lennon, M.A. (2004). Effective use of fluorides for the prevention of dental caries in the 21st century: The WHO approach. *Community Dent. Oral Epidemiol.*, 32, 319–321.
2. WHO (2002). Fluorides, Environmental Health Criteria Number, 227. Geneva, Switzerland: World Health Organization.
3. Susheela, A.K. (2003). *A Treatise on Fluorosis*, 2nd revised edn, pp. 13–14. New Delhi, India: Fluorosis Research and Rural Development Foundation.
4. Shellis, R.P. and Duckworth, R.M. (1994). Studies on the cariostatic mechanisms of fluoride. *Int. Dent. J.*, 44, 263–273.
5. Featherstone, J.D. (1999). Prevention and reversal of dental caries: Role of low level fluoride, *Community Dent. Oral Epidemiol.* 27, 31–40.
6. Ayoob, S. and Gupta, A.K. (2006). Fluoride in drinking water: A review on the status and stress effects. *Cri. Rev. Environ. Sci. Technol.*, 36, 433–487.
7. Lennon, M., Whelton, H., O'Mullane, D. and Ekstrand, J. (2005). Nutrients in drinking water, Chapter 14. In: *Fluoride*, Geneva, Switzerland: WHO, 2005. http://www.who.int/water_sanitation_health/dwq/nutrientschap14.pdf.
8. Diesendorf, M. (1995). How science can illuminate ethical debates: A case study on water fluoridation. *Fluoride*, 28, 87–104.
9. WHO (2004). Fluoride in drinking-water, Background document for preparation of WHO Guidelines for drinking water quality. Geneva, Switzerland: World Health Organization. http://www.who.int/water sanitation health /dwq/guidelines/en.
10. IAOMT (2003). Report on Policy position on ingested fluoride and fluoridation. In: David, C.K., (Ed.), *International Academy of Oral Medicine and Toxicology, Orlando.* http://www.iaomt.org/documents/IAOMT_Fluoridation _Position.pdf.

11. MRC (2002). Working Group Report: Water fluoridation and health, MRC: 47. London: Medical Research Council. http://www.mrc.ac.uk/pdf-publication-swaterfluoridation report.pdf.
12. WHO (2006). *Fluoride in Drinking Water.* Fawell, J., Bailey, K., Chilton, J., Dahi, E., Fewtrell, L. and Magara, Y. (Eds.), pp. 41–75. London, UK: IWA Publishing.
13. Shortt, H.E., Pandit, C.G. and Raghavachari, T.N.S. (1937). Endemic fluorosis in the Nellore District of South India. *Indian Med. Gaz.*, 72, 396.
14. McKay, F.S. (1928). Relation of mottled enamel to caries. *J. Am. Dent. A.*, 15, 1429–1437.
15. McKay, F.S. and Black, G.V. (1916). An investigation of mottled teeth: An endemic developmental imperfection of the enamel of the teeth, heretofore unknown in the literature of dentistry. *Dent. Cosmos.*, 58, 477–484.
16. Susheela, A.K. and Bhatnagar, M. (1999). Structural aberrations in fluorosed human teeth: Biochemical and scanning electron microscopic studies. *Curr. Sci.*, 77, 1677–1680.
17. Chilvers, C. (1983). Cancer mortality and fluoridation of water supplies in 35 U.S. cities. *Int. J. Epidemiol.*, 12, 397–404.
18. Whitford, G.M. (1997). Determinants and mechanisms of enamel fluorosis. *Ciba Found. Symp.*, 205, 226–245.
19. Dean, H.T. (1934). Classification of mottled enamel diagnosis. *J. Am. Dent. Assoc.*, 21, 1421–1426.
20. Viswanathan, G., Jaswanth, A., Gopalakrishnan, S., ilango, S.S. and Aditya, G. (2009). Determining the optimal fluoride concentration in drinking water for fluoride endemic regions in South India. *Sci. Total Environ.*, 407, 5298–5307.
21. WHO (1994). Fluorides and oral health. Report of a WHO Expert Committee on Oral Health Status and Fluoride Use. WHO Technical Report Series 846. Geneva: World Health Organization.
22. Aoba, T. (1997). The effect of fluoride on apatite structure and growth. *Crit. Rev. Oral Biol. Med.*, 8, 136–153.
23. EPA (1985). National primary drinking water regulations: Fluoride, final rule. *Federal Register*, 50, 47142–47155.
24. McKnight, C.B., Levy, S.M., Cooper, S.E. and Jakobsen, J.R. (1998). A pilot study of esthetic perceptions of dental fluorosis vs. selected other dental conditions. *ASDC J. Dent. Child.*, 65, 233–238.
25. UNICEF. Photo essay, UNICEF/India/2006/Ruhani kaur, fluorosis-mitigating the scourge. http://www.unicef.org/india/1425.html
26. Choubisa, S.L., Sompura, K., Bhatt, S.K., Choubisa, D.K., Pandya, H., Joshi, S.C. and Choubisa, L. (1996). Prevalence of fluorosis in some villages of Dungarpur district of Rajasthan. *Indian J. Environ. Health*, 38, 119–126.
27. McDonagh, M., Whiting, P., Bradley, M., Cooper, J., Sutton, A., Chestnutt, I., Misso, K., Wilson, P., Treasure, E. and Kleijne, J. (2000). A Systematic Review of Public Water Fluoridation, NHS Centre for Reviews and Dissemination, University of York, York YO10 5DD, 2000. http://www.nhs.uk/conditions/fluoride/documents/crdreport18.pdf.
28. Xianga, Q., Zhoua, M., Wua, M., Zhoub, X., Linb, L., Huangb, J. and Liangc, Y. (2009). Relationships between daily total fluoride intake and dental fluorosis and dental caries, *JNMU*, 23, 33–39.

29. Jha, S.K., Nayak, A.K. and Sharma, Y.K. (2011). Site specific toxicological risk from fluoride exposure through ingestion of vegetables and cereal crops in Unnao district, Uttar Pradesh, India. *Ecotoxicol. Environ. Saf.*, 74, 940–946.
30. Arvind, B.A., Isaac, A., Murthy, N.S., Shivaraj, N.S., Suryanarayana, S.P. and Pruthvishet, S. (2012). Prevalence and severity of dental fluorosis and genu valgum among school children in rural field practice area of a medical college. *Asian Pac J. Trop. Dis.*, 465–469.
31. Choubisa, S.L. (2001). Endemic fluorosis in southern Rajasthan, India, Research Report. *Fluoride*, 34, 61–70.

4

Skeletal Fluorosis

4.1 Introduction

The most consistent and the best characterized toxic response to fluoride is its effect on the human skeleton, as 99% of the ingested fluoride in the human body gets stored in bones. The highly vascularized soft tissues and blood store the remaining fluoride. Once absorbed, the fluoride gets readily accommodated in the active, growing, and cancellous areas than in compact regions. The concentration of fluoride stored in various bones of the same skeleton differs with the type of bones. The pelvis accumulates higher fluoride than the limb bones; young and cancellous bones are more receptive to fluoride than old or cortical bones. Though factors such as sex, age, and type and specific part of the bone influence the concentration of fluoride in bones, fluoride accumulation gets slower with age and reaches an "equilibrium" effect after about 50 years of age.[1,2]

4.2 Action of Fluoride on Bone

The chemical composition and the physical structure of human bones get altered with ingested fluoride. The resorption and accretion of bone tissue are altered by fluoride intake, which in turn affects the homeostasis of bone mineral metabolism. A combination of osteomalacia, osteosclerosis, and osteoporosis of different degrees illustrates the bone lesions. Fluoride toxicity in bones imparts impaired bone collagen synthesis, increased metabolic turnover, and increased avidity for calcium. Fluoride toxicity is also reflected in the structural changes of bones, namely, "increased bone mass and density, exostosis (bony outgrowth) at bone surfaces, increased osteoid seam and resorption surfaces, increased osteon diameter and mottling of osteons, increased trabecular bone volume, cortical porosity and development of cartilaginous lesions in the cancellous bones."[3,4] Fluoride can replace the hydroxyl and bicarbonate ions that are associated with hydroxyapatite (mineral phase during formation of bone), forming hydroxyfluorapatite and altering the mineral structure of bones. Thus, the fluoride ions occupying the plane of the calcium ions form structurally compact and electrostatically stable structures that may shift the mineralization profile to higher levels of density and hardness. Thus, the remineralization process increases bone density, making denser and harder bones. However, high-dose administration of fluoride over a long period reduces the mechanical strength of bones.[5]

The interface between collagen and the mineral is one of the causative factors inducing bone strength. During overexposure for a long term, a rapid exchange mechanism becomes active between fluoride and hydroxyl ions in the hydroxyl apatite structure of the bone. This process of exposure may continue irreversibly. Due to this activity, the rate of synthesis of bone material (hydroxyl apatite) gets considerably increased. As a result, the development of osteosclerosis (hardening and calcifying of the bones) takes place. This denotes one of the basic symptoms of people suffering from skeletal fluorosis. The mechanical properties of bones get drastically influenced by the increasing fluoride dose into bones. This increase in bone fluoride may affect the bone matrix proteins and mineral–organic interfacial bonding; it may "interfere with bone crystal growth inhibition on the crystallite faces as well as bonding between the mineral and organic interface." It was pointed out that long-term fluoride exposure and its cumulative accumulation in the bones produce "heavier and brittle" bones that are more fragile than normal bones. However, fluoride has been used for the treatment of osteoporosis, as it stimulates the formation of bones. Further, fluoride may add to bone mass, thereby inhibiting bone resorption. However, it has been established recently that the "beneficial increase in spinal bone mass is at the expense of an increasing risk of hip fractures."[4] It was suggested that though "fluoride is the single most effective agent for increasing axial bone volume in the osteoporotic skeleton, its therapeutic window is very narrow."[4,6–8]

4.3 Fluoride Exposure Level and Skeletal Fracture

Effects on the skeleton are the best indicators of the toxic responses to fluoride and are considered to have direct public health relevance. However, studies on the association between exposure to fluoride and incidence of hip fractures are either contradicting or inconclusive. Many studies ruled out a correlation between fluoridated water and prevalence of hip fractures, whereas some studies reported otherwise. The increase in serum alkaline phosphatase due to fluoride uptake, referred to as a sign of osteoblast activity in conventional medicine, is actually a reflection of increased mortality of osteocytes in bones. After analyzing 29 studies (among these, 18 investigations provided data on hip fracture) dealing with bone fracture and bone development problems, "any association between water fluoridation and bone or hip fracture incidence was clearly ruled out."[4,9]

4.4 Skeletal Fluorosis

Due to prolonged accumulation of fluoride, bones may turn fragile and exhibit low tensile strength. Such a condition of the bone is referred to as *skeletal fluorosis*. This sorry state of affairs is not easily recognizable until it reaches the advanced stage. Though in the initial stages it may show symptoms of arthritis, it may turn out to be a crippling disability in the most advanced stages. The restrictions in spine movement could be regarded as an early warning sign of skeletal fluorosis. Symptoms of ill effects are detected in the knee, spine, neck, pelvi and shoulder joints.[3] There is a steady increase in stiffness until the entire spine becomes one continuous column of bone, a condition referred to as *poker back*. This stage refers to a condition in which the ligaments of the spine become ossified and calcified. Further, the stiffness associated with the spine rapidly gets transmitted to assorted joints in the limbs. The attachment of the ribs steadily reduces the progress of the chest during breathing. As a result, the "chest assumes a barrel shape." It is reported that there is an increasing immobilization of joints due to contractures. This induces flexion deformities at knees, hips, and other associated joints, forcing the patient to be confined to bed. Despite the fact that the entire bone structure has become affected, the mental faculties remain unimpaired until the last stage.[4,10]

The occurrence of endemic skeletal fluorosis has been reported mainly from the fluorosis endemic regions of the developing world (Figures 4.1 and 4.2). Increased risk of bone effects at total intakes higher than 6.0 mg fluoride/day and at a fluoride concentration of 1.4 mg/L are reported from India and China.[1] Skeletal fluorosis was reported from India at an average fluoride concentration as low as 0.7 mg/L that was within the range of 0.4–1.4 mg/L. Evidence of skeletal fluorosis with severe clinical manifestations was reported from areas with a concentration range of 0.7–2.5 mg/L.[11] The bone ash fluoride concentrations range from 500–1500 mg/L among people exposed to fluoride levels of 1.0 mg/L. Symptoms such as joint stiffness, sporadic pain, and osteosclerosis of the pelvis are reported at concentrations of 6000–7000 mg/L. These symptoms may turn out to be increased osteosclerosis, chronic joint pain, and moderate calcification of ligaments at concentrations of 7500–9000 mg/L. The most acute form of fluorosis (crippling fluorosis) may be observed for fluoride bone concentrations greater than 10,000 mg/L.[4] Although most of the ingested fluoride gets excreted through urine, a portion of the absorbed fluoride gets deposited in skeletal tissues. More than half of the ingested fluoride in humans is excreted through the

FIGURE 4.1
Children from Jhabua district of Madhya Pradesh suffering from skeletal fluorosis. (From UNICEF, Photo essay, UNICEF/India/2006/Ruhani kaur, fluorosis-mitigating the scourge, http://www.unicef.org/india/1425.html.)

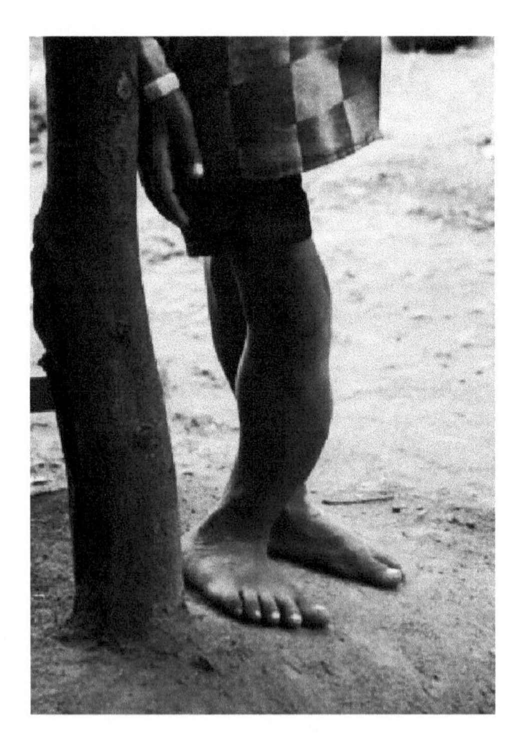

FIGURE 4.2
A 10-year-old boy from Jhabua district of Madhya Pradesh with a deformed leg and suffering from skeletal fluorosis. (From UNICEF, Photo essay, UNICEF/India/2006/Ruhani kaur, fluorosis-mitigating the scourge, http://www.unicef.org/india/1425.html.)

most crucial organ, namely the kidney. However, any failure or malfunction of the kidneys may reduce this excretion of fluoride, resulting in cumulative fluoride deposition in serum and bones. Renal failure may increase skeletal fluoride content nearly four times. This may also invite more risk of spontaneous bone fractures and possibilities of skeletal fluorosis, even at low concentrations.

4.4.1 Crippling Skeletal Fluorosis

Crippling skeletal fluorosis (Figure 4.3) is a significant cause of morbidity in a number of regions of the world. This stage is a cumulative outcome of the most advanced and severe form of skeletal fluorosis. Long-term exposure to high levels of fluoride intake coupled with malnutrition, strenuous manual labor, and impaired renal function leads to this stage. It is reported that "crippling skeletal fluorosis is marked by kyphosis (abnormally increased convexity in the curvature of the thoracic spine as viewed from the side), scoliosis (lateral curvature of the vertebral column), flexon deformity (the act of bending or the condition of being bent) of knee joints, paraplegia (paralysis of the lower part of the body including lugs) and quadriplegia (paralysis of all the four limbs)."[4] Further, the "pressure caused by osteophytes (bony outgrowth) or narrowing of intervertebral foramen and increase in size of the body of the vertebrae or narrowing of the spinal canal" leads to paralysis.[3,4] The two important forms of skeletal fluorosis observed in India are termed *genu valgum* (knock-knees) and *genu varum* (bow legs). Genu valgum, the most acute form of skeletal fluorosis, has been observed with osteosclerosis of the spine and associated osteoporosis of the limb bones. Patients suffering from fluorosis usually experience difficulty in walking because of the progressive weakness in the lower limbs. Neurological disabilities may also occur in some cases along with the spreading of this weakness to upper limbs. Since crippling fluorosis

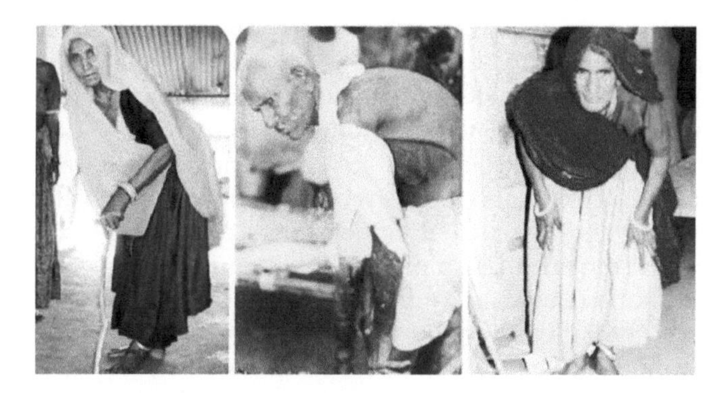

FIGURE 4.3
People in the state of Rajasthan, India, suffering from crippling skeletal fluorosis. (From Hussain, J., Hussain, I. and Sharma, K.C., *Environ. Monit. Assess.*, 162, 1–14, 2010.)

develops slowly and relentlessly, the neurological deficits may cause a slight trauma. The co existence of skeletal crippling deformities at the knee, hips, and other joints sends a confusing signal in diagnosis. It is difficult to ascertain whether such disabilities are actually induced by skeletal deformities or by neurological lesions. Such cases present a "wide spectrum of neurological deficits, which may be found manifesting either the lower motor neuron or the upper motor neuron defects, or both, which is much more common."[4] This refers to a stage in which the "anatomical features of the cervical spine will be affected and in advanced cases marked cachexia may develop due to disuse of limb and trunk muscles."[4] Another noticeable feature at this stage is the occurrence of a progressive high-frequency perceptive deafness. However, total deafness is a remote possibility. It was suggested that "the compression of the nerve in the sclerosed and narrowed auditory canal may account for the deafness" that is caused due to fluorosis.[4,13,14] It could be inferred that the acute forms of crippling skeletal fluorosis have severe bearings on the socio-economic aspects of human life, as they bring in issues such as loss of work and livelihoods, social aloofness, psychological trauma, and above all, the loss of the will to live.

4.5 Fluoride Level and Effects Related to Skeletal Fluorosis

On a rough estimate, a fluoride dose of 10–20 mg/day (equivalent to 5–10 mg/L of fluoride in water, for an individual ingesting 2 L/day) for at least 10 years may induce crippling skeletal fluorosis.[7] However, Indian research substantiates the prevalence of skeletal fluorosis at very low fluoride concentrations of 1.35 and 0.7 mg/L, and crippling skeletal fluorosis at or higher than 2.8 mg/L.[11,16–18] A high prevalence (51.1%) of crippling skeletal fluorosis (genu valgum) at 9–13 mg/L of fluoride has been reported from the state of Madhya Pradesh in India.[19] Interestingly, the prevalence of skeletal fluorosis for the same fluoride exposure of 3.0 mg/L was found to be different (19.6% and 42.2%) in two villages of the state of Punjab. In Rajasthan, a change in fluoride level from 2.5 to 2.6 mg/L resulted in an increase in the prevalence of skeletal fluorosis from around 7%–16%. At a fluoride level of 6 mg/L, 63% was the maximum skeletal fluorosis reported.[20] A lower prevalence of skeletal fluorosis (around 13%) for a fluoride exposure of 3.2 mg/L is reported in China, whereas comparatively higher values (around 40%–59%) are reported for typical Indian villages. In Andhra Pradesh state, a study was conducted on correlating the influence of the duration of exposure and the age of the people consuming water at fluoride concentrations of 9 mg/L. It was demonstrated that a daily dose of 36–54 mg of fluoride is ingested by consuming 4.6 L of water.[21] As a result, on residing in the village, the skeletal fluorosis could start in 10 years and can reach 100% at 20 years.[1,7] All these research findings demonstrate that the occurrence and prevalence of fluorosis can vary widely among different locations having almost similar fluoride concentrations. It means that apart from the fluoride concentration in drinking water, the prevalence of fluorosis can be affected by a number of other factors, as demonstrated in fluoride endemic Indian villages.

4.6 Significance of Other Factors

The relatively high prevalence of fluorosis in India at low fluoride levels needs attention. As cited, Indian research revelations provide ample evidence on the critical role of malnutrition and poverty in the prevalence and severity of fluorosis. Vitamin C deficiency and poor nutrition are also found to enhance the prevalence of fluorosis. The crippling skeletal fluorosis "genu valgum" is found to be more predominant in poorly nourished children with a low calcium intake. Such children in the endemic areas of fluorosis may also develop secondary hyperparathyroidism.[10,17] As cited earlier, at the same fluoride levels, a markedly different incidence of skeletal fluorosis was reported from India. The lower incidence of fluorosis was later associated with higher hardness of water. Thus, it appears that calcium and magnesium components of hard water play a "protective influence" against fluorosis. In China, studies suggested that prevalence of skeletal fluorosis was found to vary with the nutritional status of the habitations.[22] A study carried out in the Anantapur district of Andhra Pradesh, the worst fluoride-affected state in India, demonstrated a high content of fluoride (0.2–11.0 mg/kg) in the locally grown agricultural crops.[23] The mean fluoride content of the brewed teas was three to four times higher than the national mean of the tap water fluoride level in the United States.[24] So, it is plausible to infer that the contribution of fluoride through various food items may also add to the reported occurrence of skeletal fluorosis at low fluoride levels.[1] Thus, the fluoride content in food stuff also turns significant. However, variations in the occurrence of different forms of fluorosis are reported at the same fluoride exposure levels. These variations could be obviously attributed to the nutritional status of the people who are exposed to fluoride. Based on the research findings from China and India, a "clear excess risk" of skeletal adverse effects is predicted by the WHO for a total fluoride intake of 14 mg/day. Further, an "increased risk" of skeletal effects higher than 6 mg/day is also predicted. Recent research suggests that a fluoride level higher than 1.33 mg/L in drinking water may turn out to be a health risk. Every increase of 0.5 mg/L of water fluoride level may result in an increase in the 52 mg/kg of bone fluoride level during 2–3 years. The Agency for Toxic Substances and

Disease Registry (ATSDR) recommended an optimum fluoride dose level of 4 mg/day for adults. However, it was pointed out that the consumption of water with fluoride concentrations greater than 0.65 mg/L may be enough to cross this limit. Thus, drinking water with a fluoride concentration range of 0.5–0.65 mg/L is recommended as a "safe range" for preventing habitations of India residing in fluoride endemic regions from being placed "at risk."[25]

4.7 Recent Developments

A study conducted in China in 2012 (on 40 villages in Yuanmou County region) turns significant as it cites reduction in fluorosis over the past 20 years. From 1984 to 2006, a reduction of around 22% in dental fluorosis was reported. Further, over this period, the number of skeletal fluorosis cases was reduced from 327 to 148. This observed improvement in the reduction of fluorosis cases was credited with the supply of a low fluoride drinking water supply system, which has been successfully executed by the local governments in China over the past two decades.[26] This research output from China[26] should be an eye-opener, especially to the local governments of the fluoride endemic regions of the developing world. The high-risk areas with fluoride concentration above guideline values should be identified in every region. Dialogue between the government and local villages will help identify priority areas for intervention and budgetary allocation for fluorosis treatment. The weight of scientific evidence clearly suggests that the provision of safe drinking water supply is the only effective way for eliminating or reducing fluoride hazards. Even if safe water is provided to the inhabitants of the fluoride endemic areas, it may take a long time to eliminate at least the visible impacts of fluorosis in humans. No doubt, groundwater may remain the primary option for drinking in all the fluoride endemic remote villages where safe drinking water is still a distant dream. The ever-increasing demands of water for all other activities should also be properly accounted for when planning for alternative water sources. If fluoride testing is not carried out, the common practice of digging wells for additional water supply may increase the health risks. Reliable facilities available at a reasonable cost for fluoride tests in the fluoride endemic rural areas present another option that local governments should utilize in future. Surely, increased understanding and awareness of the relationship among fluoride levels, duration of exposure, and anticipated health impacts would help reduce the health risks. The addition of new areas in our fluoride maps cautions that fluorosis relief requires long-term monitoring and raising people's awareness on the effects of fluoride on human health.

4.8 Summary

- The toxic response to fluoride is mostly reflected in its effects on the human skeleton, as the bulk of the ingested fluoride in the human body gets stored in bones. Although the concentration of fluoride in bones varies with age, sex, and type and specific part of bone, its accumulation gets slower with age and reaches an equilibrium effect after about 50 years of age.
- The chemical composition and the physical structure of human bones get altered with the ingested fluoride. Fluoride toxicity in bones imparts impaired bone collagen synthesis, increased metabolic turnover, and increased avidity for calcium.
- The increased fluoride dose into bones reduces their mechanical properties. Overexposure to fluoride and its cumulative build-up in the bones makes bones "heavier and brittle."
- It has been suggested that the valuable increase in spinal bone mass may induce the risk of hip fractures. However, most of the studies on the relationship between fluoride exposure and the occurrence of hip fractures are either contradicting or inconclusive.
- Due to prolonged accumulation of fluoride, bones may turn fragile with a low tensile strength. Such a condition of the bone is referred to as *skeletal fluorosis* and is not easily recognizable until it reaches the advanced stage.
- An increased risk of bone effects at total intakes higher than 6.0 mg fluoride/day and at a fluoride concentration of 1.4 mg/L are reported from India and China. Skeletal fluorosis was reported from India at an average fluoride concentration as low as 0.7 mg/L that is within the range of 0.4–1.4 mg/L.
- It could be inferred that the crippling skeletal fluorosis is an important ground of morbidity in many fluoride endemic regions of the world. This stage is a cumulative outcome of the most advanced and severe form of skeletal fluorosis.
- The two important forms of skeletal fluorosis observed in India are termed *genu valgum* (knock-knees) and *genu varum* (bow legs). Genu valgum, the most acute form of skeletal fluorosis, has been observed with osteosclerosis of the spine and associated osteoporosis of the limb bones.
- Skeletal and crippling skeletal fluorosis persuades sharp social impacts on issues such as psychological shock, loss of work and employment, social remoteness, and above all, the loss of the will to live.

- As a rough estimate, a fluoride dose of 10–20 mg/day for at least 10 years may induce crippling skeletal fluorosis. However, Indian research substantiates the prevalence of skeletal fluorosis at very low fluoride concentrations of 1.35 and 0.7 mg/L and of crippling skeletal fluorosis at or higher than 2.8 mg/L.
- The occurrence, prevalence, intensity, and severity of fluorosis may vary widely among different regions and habitations having similar fluoride concentrations. Apart from the fluoride concentration in drinking water, the prevalence of fluorosis can be affected by a number of other factors.
- The crippling skeletal fluorosis genu valgum is found to be more predominant in poorly nourished children with a low calcium intake. Such children in the endemic areas of fluorosis may also develop secondary hyperparathyroidism.
- The ATSDR recommended an optimum fluoride dose level of 4 mg/day for adults. However, the consumption of water with fluoride concentrations greater than 0.65 mg/L may be enough to cross this limit.
- Drinking water with a fluoride concentration range of 0.5–0.65 mg/L is recommended as a "safe range" for preventing habitations of India residing in fluoride endemic regions from being placed "at risk."
- A fluoride level higher than 1.33 mg/L in drinking water may turn out to be a health risk. Every increase of 0.5 mg/L of water fluoride level may result in an increase of 52 mg/kg of bone fluoride level within a few years.

References

1. WHO (2002). Environmental Health Criteria Number, 227, Fluorides. Geneva, Switzerland: World Health Organization.
2. WHO (1970). Fluorides and Human Health, Monograph Series, 59, 364. Geneva, Switzerland: World Health Organization.
3. Susheela, A.K. (2003). *A Treatise on Fluorosis.* 2nd revised edn. New Delhi, India: Fluorosis Research and Rural Development Foundation.
4. Ayoob, S. and Gupta, A.K. (2006). Fluoride in drinking water: A review on the status and stress effects. *Cri. Rev. Env. Sci. Technol.*, 36, 433–487.
5. Chachra, D., Turner, C.H., Dunipace, A.J. and Grynpas, M.D. (1999). The effect of fluoride treatment on bone mineral in rabbits. *Calcif. Tissue Int.*, 64, 345–351.
6. Diesendorf, M. (1996). Fluoridation: Breaking the silence barrier. In: Brian, M., (Ed.), *Confronting the Experts*, pp. 45–75. New York: State University of New York Press.

7. ATSDR (2003). Report on Toxicological Profile For Fluorides, Hydrogen Fluoride and Fluorine. Atlanta, GA: U.S. Department of Health and Human Services, Public Health Service Agency for Toxic Substances and Disease Registry.
8. Dequeker, J. and Declerck, K. (1993). Fluor in the treatment of osteoporosis: An overview of thirty years clinical research, Department of Rheumatology and Arthritis, K.U. Leuven, U.Z. Pellenberg, Belgium. *Schweiz. Med. Wochenschr.*, 123, 2228–2234.
9. McDonagh, M., Whiting, P., Bradley, M., Cooper, J., Sutton, A., Chestnutt, I., Misso, K., Wilson, P., Treasure, E. and Kleijnen, J. (2000). A Systematic Review of Public Water Fluoridation. York, UK: NHS Centre for Reviews and Dissemination, University of York. http://fluoride.oralhealth.org/papers/pdf /yorkreport.pdf.
10. Reddy, D.R. and Deme, S.R. (2000). Skeletal Fluorosis, http://210.210.19.130 /vmu1.2/dmr/dmrdata/cme/fluorosis/Fluorosis.htm.
11. Diesendorf, M. (1991). The health hazards of fluoridation: A re-examination. *Int. Clin. Nutr. Rev.*, 10, 304–321.
12. UNICEF. Photo essay, UNICEF/India/2006/Ruhani kaur, fluorosis-mitigating the scourge. http://www.unicef.org/india/1425.html.
13. Reddy, D.R., Ravishankar, E.V., Prasad, V.S. and Muralidhar, B.N. (1994). Minor trauma causing major cervical myelopathy in fluorosis, ICRAN 94. *Gold Coast*, 585–586.
14. Rao, A.B.N. and Siddiqui, A.H. (1962). Some observations on eighth nerve function in fluorosis. *J. Laryng. Otol.*, 74, 94–99.
15. Hussain, J., Hussain, I. and Sharma, K.C. (2010). Fluoride and health hazards: Community perception in a fluorotic area of central Rajasthan (India): An arid environment. *Environ. Monit. Assess.*, 162, 1–14.
16. Jolly, S.S, Singh, B.M., Mathur, O.C. and Malhotra, K.C. (1968). Epidemiological, clinical and biochemical study of endemic dental and skeletal fluorosis in Punjab. *Brit. Med. J.*, 4, 427–429.
17. Krishnamachari, K.A. (1976). Further observations on the syndrome of endemic genu valgum of South India. *Ind. J. Med. Res.*, 64, 284–291.
18. Choubisa, S.L. (2001). Endemic fluorosis in southern Rajasthan, India, Research Report. *Fluoride*, 34, 61–70.
19. Chakma, T., Rao, P.V., Singh, S.B. and Tiwary R.S. (2000). Endemic Genu valgum and other bone deformities in two villages of Mandla district in Central India. *Fluoride*, 33, 187–195.
20. Choubisa, S.L., Choubisa, D.K., Joshi, S.C. and Choubisa, L. (1997). Fluorosis in some tribal villages of Dungapur district of Rajasthan, India. *Fluoride*, 30, 223–228.
21. Saralakumari, D. and Rao, R.P. (1993). Endemic fluorosis in the village Ralla Ananthapuram in Andhra Pradesh, an epidemiological study. *Fluoride*, 26, 177–180.
22. Chaoke, L., Rongdi, J. and Shouren, C. (1997). Epidemiological analysis of endemic fluorosis in China. *Environ. Carcinogen. Ecotoxicol. Rev.*, 15, 123–138.
23. Rao, V and Mahajan, C.L (1991). Fluoride content of some common south Indian foods and their contribution to fluorosis, *J. Sci. Food. Agric.*, 51, 275–279.
24. Pehrsson, P.R., Patterson, K.Y. and Perry, C.R. (2011). The fluoride content of select brewed and microwave-brewed black teas in the United States. *J. Food Compos. Anal.*, 24, 971–975.

25. Viswanathan, G., Jaswanth, A., Gopalakrishnan, S., Siva Ilango, S. and Aditya, G. (2009). Determining the optimal fluoride concentration in drinking water for fluoride endemic regions in South India, *Sci. Total Environ.*, 407, 5298–5307.
26. Chen, H.F., Yan, M., Yang, X.F., Chen, Z., Wang, G.A., Schmidt-Vogt, D., Xu, Y.F. and Xu, J.C. (2012). Spatial distribution and temporal variation of high fluoride contents in groundwater and prevalence of fluorosis in humans in Yuanmou County, Southwest China. *J. Hazard. Mater.*, 235–236, 201–209.

5

Stress Effects of Fluoride on Humans

5.1 Introduction

Clinical and epidemiological studies related with human health or stress effects denote important sources of data. However, the crux of the fluoride-related problems relies on the extent of coverage of the affected or sensitive subpopulations. This is significant for deriving conclusions from the toxicological viewpoint.[1] Since the range of safety is frequently unknown, clinical studies fail to identify effect levels. Thus, when such data are extensively used, it may be impossible to obtain exceptionally rigorous guideline values based on the application of unsuitable uncertainty parameters or factors. However, clinical studies and epidemiological observations habitually constitute a precious provider that is used for assigning a weight of evidence for a meticulous approach.

5.2 Nonskeletal Fluorosis

The interaction of fluoride with soft tissues, organs, and other systems of the human body induces nonskeletal fluorosis. As a result, the skeletal muscles, erythrocytes, gastrointestinal mucosa, ligaments, spermatozoa, and thyroid glands of humans will be either affected or damaged. Destruction of actin and myosin filaments in the muscle tissues reduces muscle energy. Due to this muscle weakness and corresponding loss of muscle energy, fluorosed patients find themselves unfit for normal routine activities.[2]

5.3 Fluoride and Cancer

Numerous epidemiological and experimental studies conducted world over raised concern on the impact of fluoride in drinking water and morbidity or mortality due to cancer. Though positive correlations were reported by some studies that cited statistically significant associations between fluoridation index and cancer, many reviews have not accepted such findings.[3–7] Since fluoride can have a mitogenic effect on osteoblasts, it may increase the risk for osteosarcoma.[8] The WHO suggest that apprehensions in this direction "cannot be casually dismissed."[9] An ecological study conducted in 1991 with follow-up for up to 35 years of fluoridation that dealt with 125,000 cases of incident cancers and 2.3 million cancer deaths also ruled out any correlation. In 1999, while summarizing significant research, the Centers for Disease Control and Prevention (CDC) inferred that studies to date have produced "no credible evidence of any correlation between fluoridated drinking water and increased risk for cancer."[10] In a systematic review conducted in 2000, while considering 26 studies that explored the association between cancer incidence and water fluoride exposure, no statistically significant association was found to exist between water fluoridation and incidence of cancer.[11] While summarizing considerable research in this direction, the WHO concluded that "the weight of evidence" does not support the hypothesis that "fluoride causes cancer in humans." Though most ecological studies rule out the hypothesis of an association, their "considerable limitations preclude firm conclusions from being drawn regarding the carcinogenicity of fluoride" in humans.[9,12,13] Osteosarcoma, a rare primary malignant bone tumor, is considered the sixth leading cancer in children younger than the age of 15. The annual incidence rate in the United States is 5.4 cases per million for men and 4.0 per million for women less than 20 years of age. Though the set of causes of osteosarcoma still mostly remain unknown, many studies suggest a possible link between fluoride uptake and increased occurrence of osteosarcoma in children and adolescents. In a study conducted in the United States, the age- and sex-adjusted osteosarcoma incidence data among youths between 5 and 19 years of age are compared with the water fluoridation level. The results of the study provide "no evidence that young males are at greater risk than females of the same age group to osteosarcoma" from fluoride in drinking water. Such studies suggest that water fluoridation may not have any influence on the development of osteosarcoma for either sex or age group during childhood and adolescence.[14] The efforts to correlate fluoridated water and incidence of cancer rates have not been fruitful due to a number of inherent challenges faced in such studies. Usually, it takes years to perhaps decades after exposure to the causal factors for cancer to be diagnosed. Further, a large diversity of cancers and their potential causal factors need more time for careful analysis. In general, studies conducted on human populations demonstrated contradicting and divergent views on this subject. There are studies that demonstrate a positive correlation between fluoride ingestion and osteosarcoma (bone cancer). Many of these are elusive, showing no strong relationships, and some depict negative correlations.[15] Fluoride ingestion might increase kidney and bladder cancer rates, as hydrogen fluoride (a caustic and potentially toxic substance) may be formed under the acid conditions of urine. Workers of cryolite processing plants experience this situation due to chronic occupational exposure to fluoride dust.[15] Though some studies established a relationship between the incidence of kidney and bladder cancer and the usage of drinking water laced with fluoride, universal scientific acceptance is yet to be established.

5.4 Fluoride and Gastrointestinal System

Fluoride ingestion produces symptoms of gastric irritation, such as nausea, vomiting, and gastric pain. Further, fluoride toxicity may produce loss of appetite, gas formation and nagging pain in the stomach, chronic diarrhea, chronic constipation, and persistent headache. Other symptoms include unusual fatigue, loss of muscle power and weakness, excessive thirst and frequent urination, depression, tingling sensation in fingers and toes, allergic manifestations, and so forth. The formation of hydrofluoric acid in the acidic environment of the stomach may create irritation of the gastric mucosa.[2,10,16] The exposure level is rated as "slight to moderate" when the fluoride concentration remains below 4 mg/L. It is suggested that at this stage, a population of less than 1% of those affected may experience signs of gastrointestinal issues and may be subjected to gastrointestinal hypersensitivities.[15,17] Kidneys are generally acknowledged as the major route for fluoride excretion. Fluoride induces fatal chronic kidney diseases.[18] In Children, exposure to fluoride concentration levels higher than 2.0 mg/L may induce damage to liver and kidney functions.[19] Chronic ingestion of fluoride can have noncarcinogenic effects on kidneys. Hospital admission rates for urolithiasis (kidney stones) were reported to be higher in areas of higher fluoride concentrations. In India, patients with clear signs of skeletal fluorosis living in areas of high fluoride concentrations (3.5–4.9 mg/L) were reported to be 4.6 times more likely to develop kidney stones.[20] Urinary fluoride can be useful in public health and epidemiological studies for marking fluoride exposure and intake. High levels of fluoride in urine and serum are observed in fluorotic patients. Around 30%–50% of fluoride is excreted from urine in children and is a reflection of the total fluoride intake from multiple sources. Studies conducted on short- and long-term fluctuating patterns of urinary fluoride concentration after fluoride ingestion demonstrated that higher fluoride levels in urine are associated with higher fluoride exposure.[21]

Fluoride has an impact on the thyroid-stimulating hormone production and may affect the functions of the thyroid gland.[2] Fluoride can also bind with serum calcium, thus probably reducing myocardial contractility and inviting cardiovascular collapses. The nonulcer dyspeptic symptoms, prevalence of still and deformed childbirths are reported among fluorotic patients in endemic areas of India. The established association between increasing fluoride concentrations and decreasing birth rates is suggestive of the reproductive properties of fluoride. Though the exact reason was not fully elucidated, high fluoride ingestion may have an impact on males, including the morphology and mobility of sperm, or the levels of testosterone, follicle-stimulating hormones, and inhibin-B.[15] A considerable reduction in serum testosterone levels in people diagnosed with skeletal fluorosis was observed in India.[22] Fluoride can cause pathological changes such as lipid peroxidation and DNA damage in humans,[23] which may affect our immune system. Though some genotoxic effects cannot be excluded, the overall evidence suggests that fluoride is neither genotoxic nor allergetic in humans.[24]

5.5 Other Health Effects

Fluorine can cause functional and biochemical changes in the nervous system during pregnancy, as it is capable of crossing the blood–brain barrier and gets accumulated in the brain tissue prior to birth. Such accumulations of fluoride in the tissues of the brain will disrupt the synthesis of certain receptors and neurotransmitters in the cells of the nervous system. Fluoride has a specific effect on protein synthesis in the brain, entailing degenerative changes in the neurons, varying the degrees of loss of gray matter, and causing changes in Purkinje cells in the cerebella cortex. These changes indicate that fluoride can delay cell growth and division in the cortex and that the reduced number of mitochondria, microtubules, and vesicles in the synaptic terminal may reduce efficiency in neuronal connections. Such changes might account for some of the neurological alterations present in patients with skeletal fluorosis.[25,26] Many epidemiological studies have consistently demonstrated a reduction in children's intellectual ability (IQ) due to their excessive exposure to fluoride in drinking water. Meta-analyses dealing with the effect of fluoride exposure in drinking water on the intelligence of children also displayed a strong negative relationship between fluoride exposure and kids' IQ performance. Kidneys, being an active site of metabolism, excrete considerable fluoride (around 80%) that is ingested through drinking water and other sources. So, it is plausible that fluoride concentrations in human urine can systematically reflect the burden of fluoride coverage in drinking water as an internal exposure index.[27] Evidence developed by studies conducted in high fluoride exposure areas revealed that urine fluoride concentrations were positively associated with dental fluorosis. Many literatures have shown that exposure to high levels of fluoride in drinking water is associated with deficits in children's intelligence.[28,29] It is known that the biochemical changes induced by fluoride in proteins and associated enzymatic systems may interfere with normal brain function. This may lead to impaired cognition and memory. Fluoride may adversely affect the reaction response times and visuospatial capabilities, which will get manifested as reduced IQ scores in time-sensitive tests.[30] Compared with normal children, the chance of those with severe dental fluorosis getting lower IQ scores is the most likely. This clearly indicates that exposure to high levels of fluoride in drinking water has a negative impact on the dental health and intelligence of children. Dose–response relationships exist between urine fluoride concentrations and IQ scores as well as between fluoride levels and dental fluorosis. Even a small decline in IQ scores or dental fluorosis can induce a profound influence on individuals.[27]

Studies suggest abnormalities in the ECG of people living in fluoride endemic areas that are afflicted with skeletal and dental fluorosis. The elastic properties of the ascending aorta of such patients are found to be damaged. Fluorosis is found to have an impact on the cardiovascular system, including the heart and major vessels originating from the heart. It is also observed that fluorosis patients have left ventricular diastolic and global dysfunctions. Acute fluoride toxicity will also induce enhanced oxidative stress in human beings. Acute or cronic stage of fluorosis may produce reactive oxygen which induces severe damage or even result in the death of myocardial cells. These impacts in these cells invite issues related with left ventricular diastolic.[31]

The study conducted in 2013 in Zhaozhou county from Heilongjiang province in China confirmed the relationship between excess fluoride intake and essential hypertension in adults. It was also demonstrated that high levels of fluoride exposure in drinking water increase plasma ET-1 levels in people living in fluoride endemic areas.[32] Of late, many scientific studies clearly underlined the fact that fluoride is capable of inducing oxidative stress. Further, it may transform intracellular redox homeostasis, protein carbonyl content, and oxidative degradation of lipids; amend gene expression; and originate apoptosis. Genes related with metabolic enzymes, cell cycle, stress response, signal transduction, and cell-to-cell communication are modulated by the presence of fluoride. Biologically relevant concentrations of fluoride are capable of increasing cell migration in tumor cells and of stimulating tumor invasion.[33] Fluoride causes disturbances in the lipid metabolism in the blood of patients who are afflicted with skeletal fluorosis. Due to the inhibition of lipid synthesis by fluoride, the cholesterol content (both high-density lipoprotein [HDL] and low-density lipoprotein [LDL]) gets reduced.[34]

5.6 Summary

- The interaction of fluoride with soft tissues, organs, and other systems of the human body induces nonskeletal fluorosis. As a result, the skeletal muscles, erythrocytes, gastrointestinal mucosa, ligaments, spermatozoa, and thyroid glands of humans will be either affected or damaged.
- Though the set of causes of osteosarcoma still mostly remain unknown, many studies suggest a possible link between fluoride uptake and increased occurrence of osteosarcoma in children and

adolescents. Though most of the ecological studies rules out the hypothesis of an association, their considerable limitations preclude firm conclusions from being drawn regarding the carcinogenicity of fluoride in humans.

- Though some studies established an association between fluoridated drinking water and the incidence of kidney and bladder cancer, universal scientific acceptance is yet to be established.
- Fluoride ingestion produces symptoms of gastric irritation, such as nausea, vomiting, and gastric pain. Chronic ingestion of fluoride can have noncarcinogenic effects on the kidneys. People with clear signs of skeletal fluorosis living in areas of high fluoride concentrations are at a high risk of developing kidney stones.
- It is plausible that fluoride concentrations in human urine can systematically reflect the burden of fluoride coverage in drinking water as an internal exposure index. Evidence developed by studies in high fluoride exposure areas revealed that urine fluoride concentrations were positively associated with dental fluorosis.
- Fluoride may adversely affect the reaction response times and visuospatial capabilities, which will get manifested as reduced IQ scores in time-sensitive tests. This clearly indicates that exposure to high levels of fluoride in drinking water has a negative impact on the dental health and intelligence of children.
- Acute fluoride toxicity will also induce enhanced oxidative stress in human beings.
- Fluoride causes disturbances in the lipid metabolism in the blood of patients who are afflicted with skeletal fluorosis. Due to the inhibition of lipid synthesis by fluoride, the cholesterol content gets reduced.

References

1. WHO. (2009). World Health Organization Guidelines for Drinking-Water Quality Policies and Procedures Used in Updating the WHO Guidelines for Drinking Water Quality Public Health. WHO/HSE/WSH/09.05. Geneva, Switzerland: Public Health and the Environment, WHO Press.
2. Susheela, A.K. (2003). *A Treatise on Fluorosis*, Revised 2nd edn. New Delhi, India: Fluorosis Research and Rural Development Foundation.
3. Cohn, P.D. (1992). *An Epidemiologic Report on Drinking Water and Fluoridation*, pp. 1–17. Trenton, NJ: Fluoride and Osteosarcoma in New Jersey, Environmental Health Service.
4. Freni, S.C. and Gaylor, D.W. (1992). International trends in the incidence of bone cancer are not related to drinking water fluoridation. *Cancer*, 70, 611–618.

5. Takahashi, K., Akiniwa, K. and Narita, K. (2001). Regression analysis of cancer incidence rates and water fluoride in the U.S.A. based on IACR/ IARC (WHO) data (1978–1992), International Agency for Research on Cancer. *J. Epidemiol.*, 11, 170–179.
6. IARC. (1987). *IARC Monographs on Evaluation of Carcinogenic Risks of Chemicals to Humans.* Supplement 7: Overall Evaluations of Carcinogenicity: An Updating of IARC Monographs, Vols. 1–42. Lyon: International Agency for Research on Cancer.
7. Knox, E.G. (1985). Fluoridation of Water and Cancer: A Review of the Epidemiological Evidence. London, UK: Working Party on the Fluoridation of Water and Cancer. HMSO.
8. MRC. (2002). Working Group Report: Water Fluoridation and Health, MRC: 47. London: Medical Research Council, 2002. http://www.mrc.ac.uk/pdf -publicationswaterfluoridation report.pdf.
9. WHO. (2002). Fluorides, Environmental Health Criteria Number, 227. Geneva, Switzerland: World Health Organization.
10. Ayoob, S. and Gupta, A.K. (2006). Fluoride in drinking water: A review on the status and stress effects. *Crit. Rev. Environ. Sci. Technol.*, 36, 433–487.
11. McDonagh, M., Whiting, P., Bradley, M., Cooper, J., Sutton, A., Chestnutt, I., Misso, K., Wilson, P., Treasure, E. and Kleijnen, J. (2000). *A Systematic Review of Public Water Fluoridation.* York, UK: NHS Centre for Reviews and Dissemination, University of York. http://fluoride.oralhealth.org/papers/pdf/yorkreport.pdf.
12. ATSDR. (2003). Report on Toxicological Profile for Fluorides, Hydrogen Fluoride and Fluorine. Atlanta, GA: U.S. Department of Health and Human Services, Public Health Service Agency for Toxic Substances and Disease Registry.
13. IPCS. (2002). International Programme on Chemical Safety. Fluorides, Environmental Health Criteria, 227. Geneva, Switzerland: World Health Organization.
14. Levy, M. and Leclerc, B.S. (2012). Fluoride in drinking water and osteosarcoma incidence rates in the continental United States among children and adolescents. *Cancer Epidemiol.*, 36, e83–e88.
15. Ozsvath, D.L. (2009). Fluoride and environmental health: A review. *Rev. Environ. Sci. Biotechnol.*, 8, 59–79.
16. Susheela, A.K. (1989). Fluorosis: Early warning signs and diagnostic test. *Bull. Nutr. Found. India*, 10(2).
17. Doull, J., Boekelheide, K., Farishian, B.G., Isaacson, R.L., Klotz, J.B., Kumar, J.V., Limeback, H., Poole, C., Puzas, J.E., Reed, N.M.R., Thiessen, K.M. and Webster, T.F. (2006). *Fluoride in Drinking Water: A Scientific Review of EPA's Standards,* p. 530. Committee on Fluoride in Drinking Water, Board on Environmental Studies and Toxicology, Division on Earth and Life Sciences, National Research Council of the National Academies. Washington, DC: National Academies Press. http://www.nap.edu.
18. Chandrajith, R., Dissanayake, C.B., Ariyarathna, T., Herath, H.M. and Padmasiri, J.P. (2011). Dose-dependent Na and Ca in fluoride-rich drinking water-Another major cause of chronic renal failure in tropical arid regions. *Sci. Total Environ.*, 409, 671–675.
19. Xiong, X., Liu, J., He, W., Xia, T., He, P., Chen, X., Yang, K. and Wang, A. (2007). Dose-effect relationship between drinking water fluoride levels and damage to liver and kidney functions in children. *Environ. Res.*, 103, 112–116.

20. Singh, P.P., Barjatiya, M.K., Dhing, S., Bhatnagar, R., Kothari, S. and Dhar, V. (2001). Evidence suggesting that high intake of fluoride provokes nephrolithiasis in tribal populations. *Urol. Res.*, 29, 238–244.
21. Liu, H.Y., Chen, J.R., Hung, H.C., Hsiao, S.Y., Huang, S.T. and Chen, H.S. (2011). Urinary fluoride concentration in children with disabilities following long-term fluoride tablet ingestion. *Res. Dev. Disabil.*, 32, 2441–2448.
22. Susheela, A.K. and Jethanandani, P. (1996). Circulating testosterone levels in skeletal fluorosis patients. *Clin. Toxicol.*, 34, 183–189.
23. Wang, A.G., Xia, T., Chu, Q.L., Zhang, M., Liu, F., Chen, X.M. and Yang, K.D. (2004). Effects of fluoride on lipid peroxidation, DNA damage and apoptosis in human embryo hepatocytes. *Biomed. Environ. Sci.*, 17, 217–222.
24. NRC. (1993). *Health Effects of Ingested Fluoride, Commission on Life Sciences*, pp. 51–72, 125–128. Washington, DC: National Academy Press. http://www.nap.edu/books/030904975X/html.
25. Valdez-Jiménez, L., Soria Fregozo, C., Miranda Beltrán, M.L., Gutiérrez Coronado, O., Pérez Vega, M.I., (2011). Effects of the fluoride on the central nervous system. *Neurología*, 26, 297–300.
26. Du, L., Wan, C., Cao, X. and Liu, J., (1992). The effect of fluorine on the developing human brain. *Chin. J. Pathol.*, 21, 218–220.
27. Ding, Y., Gao, Y., Sun, H., Han, H., Wang, W., Ji, X., Liu, X. and Sun, D. (2011). The relationships between low levels of urine fluoride on children's intelligence, dental fluorosis in endemic fluorosis areas in Hulunbuir, Inner Mongolia, China. *J. Hazard. Mater.*, 186, 1942–1946.
28. Trivedi, M.H., Verma, R.J., Chinoy, N.J., Patel, R.S. and Sathawara, N.G. (2007). Effect of high fluoride water on intelligence of school children in India. *Fluoride*, 40, 178–183.
29. Wang, S.X., Wang, Z.H., Cheng, X.T., Li, J., Sang, Z.P., Zhang, X.D., Han, L.L., Qiao, X.Y., Wu, Z.M. and Wang, Z.Q. (2007). Arsenic and fluoride exposure in drinking water: children's IQ and growth in Shanyin county, Shanxi province, China. *Environ. Health Perspect.*, 115, 643–647.
30. Calderon, J., Blenda, M., Marielena, N., Leticia, C., Deogracias, O.M., Diaz-Barriga, F. (2000). Influence of fluoride exposure on reaction time and visuospatial organization in children. *Epidemiology*, 11, S153.
31. Varol, E., Akcay, S., Ersoy, I.H., Koroglu, B.K. and Varol, S. (2010). Impact of chronic fluorosis on left ventricular diastolic and global functions. *Sci. Total Environ.*, 408, 2295–2298.
32. Sun, L., Gao, Y., Liu, H., Zhang, W., Ding, Y., Li, B., Li, M. and Sun, D. (2013). An assessment of the relationship between excess fluoride intake from drinking water and essential hypertension in adults residing in fluoride endemic areas. *Sci. Total Environ.*, 443, 864–869.
33. Mendoza-Schulz, A., Solano-Agama, C., Arreola-Mendoza, L., Reyes-Márquez, B., Barbier, O., Del Razo, L.M. and Mendoza-Garrido, M.E. (2009). The effects of fluoride on cell migration, cell proliferation, and cell metabolism in GH4C1 pituitary tumour cells. *Toxicol. Lett.*, 190, 179–186.
34. Bhardwaj, M. and Shashi, A. (2012). Dose effect relationship between high fluoride intake and biomarkers of lipid metabolism in endemic fluorosis. *Biomed Prev Nutr.*, 3, 121–127. http://dx.doi.org/10.1016/j.bionut.2012.10.006.

6

Fluoride in the Environment and Its Toxicological Effects

6.1 Introduction

Fluorine is the most electronegative and the most reactive element present in the periodic table. It is always found in the nature with the combination of some other elements because of its reactivity. Fluoride is a naturally occurring, widely distributed element, and it is found in varying amounts in minerals, rocks, gases from volcanoes, and so forth. Anthropogenic sources such as coal-fired power plants; aluminum smelters; phosphate fertilizer plants; glass, brick, and tile works; and plastics factories are also responsible for the increase in the level of the fluorine in the atmosphere.[1,2] As stated in previous chapters, human beings are benefited when the fluoride uptake is of an appropriate quantity. If the fluoride uptake is lower than the minimum amount or more than the upper limit, it can adversely affect human health. Studies have shown that apart from human beings, plants, and animals also experience toxicological issues when they are exposed to fluoride. Drinking water serves as the major source of fluoride in humans; however, other environmental sources also contribute to it.[2]

6.2 Sources of Environmental Exposure

Fluorine is present in the environment mostly as fluorides. Fluoride is released into the environment either naturally or through human activities. The natural processes that are responsible for fluoride intrusion into the environment may include weathering and dissolution of minerals, volcanic eruption, and contribution of marine aerosols.[3] Fluoride is also released into the environment by manifold human activities, for example, coal combustion; processing of water and waste from various industrial processes, including steel manufacture; aluminum, copper, and nickel production;

phosphate ore processing; phosphate fertilizer production and use; glass, brick, and ceramic manufacturing; glue and adhesive production; and so forth. Among these, industrial sources such as phosphate ore production and use, and aluminum manufacture are the major anthropogenic sources of fluoride.[4,5] Controlled addition of fluoride to a public water supply, food items, dental products, and pharmaceutical products also contribute to the release of fluoride into the environment through anthropogenic sources.[6] A number of fluoride compounds that are mostly used for industrial and commercial purposes are also potential sources of fluoride in the environment. In this direction, some important fluoride compounds are listed as follows:[2,4]

- Hydrogen fluoride is predominantly used in the production of aluminum and chlorofluorocarbons (CFCs). It is used in electronic industries, cleaning glass, brick and tiles work, tanning leather, and in commercial rust removers and pickling operation in stainless steel. It has wide application in separating uranium isotopes or as a catalyst in the petroleum industry.
- Calcium fluoride has multiple uses in the fiber glass, ceramic, welding rod, glass, and fluorescent lamps industry. It is used in blending with burned lime and dolomite in the steel industry, and it is also a fluxing agent for aluminum metallurgy.
- Sodium fluoride has wide use in fluoridation of drinking water. It is used as a preservative in glues and wood, and as an insecticide. It is also used as a flux in steel and aluminum production, as well as in glass and enamel production.
- Sulfur hexafluoride is used as a gaseous dielectric medium in the electrical industry for high-voltage circuit breakers, switch gear, various electronic components, and so forth. It has some application in the production of magnesium and aluminum.
- Fluorosilicic acid and sodium hexafluorosilicate are used for the fluoridation of drinking-water supplies.

6.3 Environmental Transport, Distribution, and Transformation

In the atmosphere, fluorides exist in gaseous or particulate form. Gaseous fluorides can be either transported over long distances by the effect of wind or atmospheric turbulence or absorbed by the atmospheric water (rain, clouds, fog, snow), forming an aerosol or a fog of aqueous hydrofluoric acid. They can be removed from the atmosphere via wet deposition. Fluoride present in the particulate form can be removed from the atmosphere and deposited on land or surface water by wet and dry deposition.[7] Most of the fluoride compounds are not expected to remain in the troposphere for long periods or to migrate to the stratosphere. However, sulfur hexafluoride can reside in the atmosphere for a period ranging from 500 years to several thousand years.[2] Several factors influence the process of the transport, and transformation of fluoride in water. Some of the important factors are pH, water hardness, and the presence of ion-exchange materials such as clays. Fluorides usually combine with aluminum during their transportation and transformation process through water cycle. In soils, the transport and transformation of fluoride are influenced by pH and the formation of complexes, predominantly those with aluminum and calcium. When the soil is slightly acidic (pH 5.5–6.5), fluoride may be adsorbed more strongly by the soil particles. Fluorides strongly bond with the soil and are not easily leached from it. Terrestrial plants may accumulate fluorides from the atmosphere through the opening of stomata and from the soil via roots. Some aquatic plants and animals can bioaccumulate the soluble fluorides from water. The quantity of fluoride uptake depends on the route of exposure, how fluoride is absorbed by the body and how quickly it is consumed and excreted.

6.4 Environmental Levels and Human Exposure

The level of fluoride in surface water such as rivers, lakes, and so forth depends on its location and distance from the emission source. Fluoride concentration in the range between 0.01 and 0.3 mg/L is found in surface water,[2] whereas sea water contains more fluoride than fresh water and it ranges between 1.2 and 1.5 mg/L.[2] Fluoride concentration in groundwater is very high in areas where the natural rocks and soil are rich in fluoride. Inorganic fluoride concentration is very high in the regions where there is geothermal or volcanic activity. Anthropogenic discharges are also one of the major sources that increase the concentration of fluoride in the environment.

The fate and transport of gaseous and particulate fluorides in the environment occurring from natural and anthropogenic sources depend on various atmospheric conditions. The transportation, distribution, transformation, and deposition of airborne fluorides are governed by meteorological conditions, particulate size, chemical reactivity, and emission strength of the source. Fluorides that are released as gaseous and particulate matter are generally deposited in the vicinity of an emission source, whereas some particulates may react with other atmospheric constituents. It has been found that the fluoride concentrations in the ambient air in areas that do not have any nearby emission sources[2] are generally less than 0.1 μg/m^3. However, even in the areas that have nearby emission sources, the concentration of levels of airborne fluoride[2] usually do not exceed 2–3 μg/m^3.

Fluoride is present in various concentrations in different types of soils. The fluoride concentrations in the areas without natural phosphate and fluoride deposits range from 20 to 1000 μg/g, whereas the concentration may increase to several thousand micrograms per gram for the soil that is enriched with natural deposits of fluoride.[2] Many foodstuffs contain very small amounts of fluoride; however, very high amounts of fluoride are reported in fish and tea leaves. The exposure of individuals to fluorides varies considerably. It depends on the level of fluoride in drinking water or dietary intake, the use of fluoridated dental/pharmaceutical products and, in some cases, the levels of fluoride in indoor air. The inhalation of airborne fluoride generally contributes a minor stake in the total fluoride (TF) intake. Infants who are fed with formula receive 50–100 times more fluoride than infants who are fed exclusively breast milk. In the case of young children, the ingestion of toothpaste containing fluoride contributes a significant amount to their TF intake. The intake of fluoride in adults is mainly through foodstuffs and drinking water. Generally, 2 mg/day of fluoride finds its way into the body of children and adolescents.[2] Although adults may have a higher absolute daily intake of fluoride in terms of milligrams, the daily fluoride intake of children, expressed on a milligram-per-kilogram body weight basis, may exceed that of adults.[2] Occupational individuals working in industries such as aluminum, fertilizer, iron, oil refining, semiconductor, phosphate, ore, and steel are more susceptible to get exposed to fluoride via inhalation or dermal contact. Recent studies[2] reported that the concentration of fluoride in the pot rooms generated from aluminum smelters is found to be in the order of 1 mg/m^3.

6.4.1 Fluoride from Dental Products

It has been established that different dental and pharmaceutical products that are used by people in their daily life for oral hygiene contain fluoride. These include toothpastes (1.0–1.5 g/kg fluoride), gels (0.25–24.0 g/kg fluoride), fluoride tablets (0.25–1.00 mg fluoride/tablet) and so forth.[6] Fluorides in the form of sodium fluoride (NaF), sodium monofluorophosphate (Na_2FPO_3) and tin fluoride (SnF_2), or in the form of different amines can be added to toothpastes. Dental products such as toothpaste and tooth powder may also contribute to the TF intake. In pharmaceutical products, fluoride can be added in the form of NaF or Na_2FPO_3. The intake of fluoride through dental products varies from person to person on the basis of their practice and exercise of cleaning teeth. Generally, small children are more susceptible to ingestion of toothpaste while cleaning their teeth. Studies on toothpaste suggest that during cleaning of the teeth, most children ingest approximately 20% of the toothpaste. It has been found that ingesting the toothpastes containing fluoride may result in a contribution of about 0.50–0.75 mg of fluoride per child per day.[7–9] Some of the researchers conducted studies to determine the fluoride concentration in dental products. Kumar and Yadav (2014) estimated the fluoride concentration in toothpastes and tooth powder available in the market of Rampur district, Uttar Pradesh, India.[10] The water-soluble fluorides in the toothpastes were found to vary from 20 to 1100 mg/kg and those in tooth powders ranged from 9.11 to 22.11 mg/kg.[10] Yadav et al. (2007) collected 15 samples of toothpaste and determined the fluoride concentration present in them.[11] It was found that one gram of toothpaste contributes approximately 53.5–338.5 µg of fluoride to the human body with a mean value of 183.78 µg.[11]

6.4.2 Fluoride from Food and Beverage

Fluorides may find their way into the human body through diet. Generally, food contains low levels of fluoride, though a trace amount of fluoride may be present in all foodstuffs. However, fluoride may be present in higher amounts in the food grown in areas where soils contain higher amounts of fluorides or where phosphate fertilizers are used for agricultural purposes. Tea and some seafood are reported to have very high levels of fluoride. A summary of various studies in which the levels of fluoride in foods have been assessed is presented in Table 6.1.

The concentration of fluoride in food items mainly depends on the nature of soil and the quality of water used for irrigation; thus, it varies from place to place. Several studies were conducted in the past to determine the intake of fluoride through the food items. Most of the studies suggested that the concentration of the fluoride in the raw food items is the function of the fluoride content in the water that is used for irrigation. Some of the studies related to the fluoride concentration in Indian food items with reference to the Indian context are interesting.

Rao and Mahajan (1990) conducted a survey in 41 villages of Anantapur district, Andhra Pradesh, India, to determine the fluoride concentration in 98 different food items.[33] In this area, a fluoride concentration up to 4.5 mg/kg was reported in irrigation water and this fluoride finds its way into most of the food items that are grown in this area. It was estimated that 32 locally grown food items generally had a fluoride concentration ranging from 0.2 to 11.0 mg/kg. The TF intake from both food and drinking water was found to be in the range of 2.2–7.3 mg/d (0.05–0.32 mg/d/kg body weight): Food contributes a large fraction to TF intake, ranging from 1.3 to 3.4 mg/d.[33] Gautam et al. (2010) estimated the concentration of fluoride in the food items collected from Nawa tehsil in Nagaur district of Rajasthan.[34] The fluoride content in water was found to be in the range of 0.92–14.62 mg/L. Leafy vegetables grown in fluoride endemic areas were found to be the most susceptible to fluoride and they contained fluoride in the range of 8.08–25.70 mg/kg. Fluoride concentrations in cereal crops were also high (ranging from 1.88 to 18.98 mg/g).[34] Bhargava and Bhardwaj (2009) conducted a study in 10 endemic villages of north-east Rajasthan to determine the fluoride levels in different local food items.[35] Food items such as vegetables, cereals, fodder, and milk were collected from the fluoride endemic villages and analyzed for fluoride content. It was reported

TABLE 6.1

Fluoride Concentration in Different Foodstuffs

Type of Food	Test Insights	Fluoride Concentration (mg/kg)[a]	Study Area	References
Milk and milk products	30 samples of milk and milk products	0.23–1.36	Connersville and the Richmond community, Indiana (USA)	Jackson et al.[12]
	42 different types and brands of milk	0.007–0.068	Houston, Texas, USA	Liu et al.[13]
	66 cow milk samples	0.043–0.147	Dindigul district, Tamil Nadu, India	Amalraj and Pius[14]
	Pasteurized milk supplied to households	0.143–0.157	United kingdom (non-fluoridated area)	Duff[15]
	Untreated milk samples	0.162–0.173	United kingdom (where the farm water supply was fluoridated at 1 mg/L level)	Duff[15]
	68 samples of market milk	0.007–0.086	Canada	Dabeka and McKenzie[16]
	Chocolate-flavored milk for infant	0.05–1.27	Bauru municipality, Brazil	Buzalaf et al.[17]
	Soy beverages for infant	0.09–0.29	Bauru municipality, Brazil	Buzalaf et al.[17]
Meat and poultry	Mechanically separated chicken and turkey	0.08–8.63	Corvallis, Oregon, USA	Fein and Cerklewski[18]
	9 kinds of deboned poultry meat	0.3–2.7	Poland	Jedra et al.[19]
	55 samples of meat and poultry	0.03–1.41	Connersville and the Richmond community, Indiana (USA)	Jackson et al.[12]
	25 ready to eat samples of meats and chicken for infant	0.01–8.38	Iowa City, Iowa	Heilman et al.[20]
Fish	Range of fluoride levels in skeletal bone of saltwater fish	45–1207	–	Camargo[21]

(Continued)

TABLE 6.1 (*Continued*)

Fluoride Concentration in Different Foodstuffs

Type of Food	Test Insights	Fluoride Concentration (mg/kg)[a]	Study Area	References
	Range of fluoride levels in muscle of saltwater fish.	1.3–26	–	Camargo[21]
	Different variety of fishes samples included in the Valencian Community Total Diet Study	2.10–11.10	Spain	Rocha et al.[22]
	3 different species of fish (The fluoride concentration in the water samples from where fishes were caught were in the range of 0.035–0.051 mg/L)	2.35–274.29	Alappuzha district, Kerala, India	Thomas and James[23]
Baked goods and cereals	129 samples of uncooked grain products	0.07–1.36	Connersville and the Richmond community, Indiana (USA)	Jackson et al.[12]
	528 staple food grain samples	1.16–4.94	Dindigul district, Tamil Nadu, India	Amalraj and Pius[14]
	66 cooked rice samples	0.34–0.73	Dindigul district, Tamil Nadu, India	Amalraj and Pius[14]
	Rice samples included in the Valencian Community Total Diet Study	2.20	Spain	Rocha et al.[22]
	9 ready to eat samples of cereals for infant	0.01–0.31	Iowa City, Iowa	Heilman et al.[20]
	Cereals for infant	0.20–7.84	Bauru municipality, Brazil	Buzalaf et al.[17]
	Biscuits for infant	0.34–13.68	Bauru municipality, Brazil	Buzalaf et al.[17]

(*Continued*)

TABLE 6.1 (*Continued*)

Fluoride Concentration in Different Foodstuffs

Type of Food	Test Insights	Fluoride Concentration (mg/kg)[a]	Study Area	References
Vegetables	65 samples of vegetables	0.38–5.37	Warsaw market, Poland	Sawilska-Rautenstrauch et al.[24]
	660 green leafy vegetable samples	0.58–7.68	Dindigul district, Tamil Nadu, India	Amalraj and Pius[14]
	Samples of various vegetable included in the Valencian Community Total Diet Study	0.74–4.76	Spain	Rocha et al.[22]
	48 ready to eat samples of vegetables for infant	0.01–0.42	Iowa City, Iowa	Heilman et al.[20]
	78 samples of uncooked vegetables	0.003–1.930	Connersville and the Richmond community, Indiana (USA)	Jackson et al.[12]
Fruits and fruit juices	Different fruit juice samples	0.07–0.53	Davangere city, India	Thippeswamy et al.[25]
	105 juice samples	0.67	Mexico City, Mexico	Jimenez-Farfan et al.[26]
	88 ready to eat samples of fruits and desserts for infant	0.01–0.49	Iowa City, Iowa	Heilman et al.[20]
	26 samples of fruits	0.01–0.84	Connersville and the Richmond community, Indiana (USA)	Jackson et al.[12]
Fats and oils	14 samples of fats and oils	0.05–0.62	Connersville and the Richmond community, Indiana (USA)	Jackson et al.[12]
Sugars and candies	15 samples of sugar and sweets	0.07–0.60	Connersville and the Richmond community, Indiana (USA)	Jackson et al.[12]
Beverages	32 samples of beverages	0.04–0.93	Connersville and the Richmond community, Indiana (USA)	Jackson et al.[12]

(*Continued*)

TABLE 6.1 (*Continued*)

Fluoride Concentration in Different Foodstuffs

Type of Food	Test Insights	Fluoride Concentration (mg/kg)[a]	Study Area	References
	12 different carbonated soft drinks	0.19–0.42	Davangere city, India.	Thippeswamy et al.[25]
	57 carbonated drinks	0.43	Mexico City, Mexico	Jimenez-Farfan et al.[26]
	332 soft drinks samples	0.02–1.28	Iowa, USA	Heilman et al.[27]
Tea	6 different kind of tea	1.97–8.64	Taiwan	Lung et al.[28]
	Different types of tea products	170–878	China	Wong et al.[29]
Bottled drinking water	10 different types of bottled drinking water	0.06–1.05	Davangere city, India	Thippeswamy et al.[25]
	20 types of bottled waters	0.21	Mexico City, Mexico	Jimenez-Farfan et al.[26]
	10 types of bottled waters recommended for use by infants and young children	0.08–0.30	Poland	Opydo-szymaczek and Opydo[30]
	15 randomly selected commercial brands (12 local brands + 3 imported brands) of bottled water	0.5–0.83	Riyadh, Saudi Arabia.	Aldrees and Al-Manea[31]
	29 commercially available brands of bottled waters	0.19–1.07	Algeria	Bengharez et al.[32]

[a] For liquid items, concentrations are in mg/L.

that the fluoride concentrations in vegetables and cereals varied, respectively, from 3.91 to 29.15 mg/kg and 0.45 to 5.98 mg/kg. Fluoride content in milk samples was found to be in the range of 0.37–6.85 mg/L. Results suggested that along with drinking water, food items, and milk also contribute to the TF intake.[35] Raghavachari et al. (2008) conducted a study on the fluoride concentration present in the food items available at Palamau district of Jharkhand, where food is grown with irrigation water that has a fluoride concentration 0.10–12.30 mg/L.[36] The fluoride content of cereals and pulses ranges from 1.5 to 1.78 mg/kg and from 1.46 to 2.28 mg/kg, respectively. The concentration of fluoride in vegetables is very low (0.14–0.23 mg/kg) compared with that of cereals and pulses. Fluoride intake through food items alone was found to be between 0.97 and 1.23 mg/capita/day.[36] Ramteke et al. (2007) performed a study on the fluoride concentration in commonly used food items such as rice, corn, wheat, and lentils (dal) in Dhar and Jhabua districts of Madhya Pradesh.[37] It was found that the fluoride concentrations in rice, corn, wheat, and lentils (dal) were in the ranges of 0.51–5.52, 10.2–40, 0.75–9.02, and 1.1–13.42 mg/kg, respectively. The TF consumption was in the range of 10.7–21.21 mg/capita/person. The maximum consumption of fluoride was 21.21 mg/day in the age group of 31–45 years.[37] Yadav et al. (2012) investigated fluoride concentration in one crop (wheat) and two vegetables (potato and tomato) that were collected from seven villages of Dausa district in Rajasthan, India.[38] The fluoride concentration in groundwater samples obtained from hand pumps and open wells of these seven villages was found to vary from 5.1 to 14.7 mg/L. The fluoride accumulation in crop (wheat) ranged from 3.24 to 14.3 mg/kg, and the fluoride content in vegetables ranged from 1.10 to 4.60 mg/kg.[38]

6.4.3 Fluoride in Soil

Fluorine concentration in soil generally varies between 150 and 400 mg/L. Soils that are derived from rocks with a high fluorine content or that are affected by anthropogenic sources contain a very high concentration (1000 g/kg) of fluorine.[39] Fluoride mobility in soil is highly dependent on the sorption capacity of the soil and varies with pH, types of sorbents, and soil salinity.[40] This is because the fluoride concentration in the soil differs from place to place. Jha (2012) studied the distribution of the fluoride in the soils of Indo-Gangetic plains.[41] The fluoride distribution in soil profiles and surface soil (0–15 cm) samples were analyzed. Results demonstrated that the TF in the profiles varied from 248 to 786 mg/kg. The $CaCl_2$ extractable soluble fluoride (FCa) was found to be in the range of 1.68–99.1 mgF/kg soil. While in surface soils, the TF and FCa ranged from 118 to 436 mg/kg and from 1.01 to 5.05 mg/kg, respectively.[41] Mishra et al. (2009) conducted an experiment to measure the effect of fluoride emission from an aluminum smelter located at Hirakud in western Orissa on the environment.[42] The study was carried out in a radius of 5 km from the plant. It was found that the concentration of fluoride in soil varied from 88.30 to 191.20 mg/L.[42] Jha et al. (2008) estimated the fluoride concentration in the soil in the vicinity of a brick field in the suburb of Lucknow, India.[43] It was observed that the water-soluble fluoride in the surface soil varied from 0.59 to 2.74 mg/L, and the $CaCl_2$ extractable fluoride ranged from 0.69 to 3.18 mg/L. The mean TF concentration in surface soil varied from 322 to 456 mg/kg.[43] Chaudhary et al. (2009) determined the fluoride concentration in the soil samples from 60 village sites in the Indira Gandhi, Bhakra, and Ganga canal catchment areas of north-west Rajasthan, India.[44] Results suggested that the mean water-extractable soil fluoride concentrations varied from 0.50 to 3.00 mg/L. It was concluded that the heavy use of diammonium phosphate (DAP) fertilizer is a possible source of elevated fluoride in the soil of that area.[44] Anbuvel et al. (2014) studied the accumulation of fluoride in the soil of eight villages near the bank of Thovalai Channel in Kanyakumari district, Tamilnadu.[45] The result showed that the fluoride content of the soil varied from 1.0 to 3.2 mg/L. Since this area is free from industrial activity, the heavy use of phosphate fertilizers over long periods may be the reason for the increased concentrations of fluoride in the soil.[45]

6.4.4 Fluoride in Tobacco and Pan Masala

In India, a large number of people are addicted to tobacco and pan masala. While calculating the TF intake, generally these items are not taken into consideration. Generally, tobacco and pan masala are not swallowed, only some fraction of them is ingested during their consumption, which ultimately becomes available to the body for absorption along with sublingually absorbed fluoride. Therefore, these items should be considered while calculating the TF intake by a human body. The exposure of fluoride through pan masala and tobacco differs from person to person depending on their consumption habits. Yadav et al. (2007) estimated the fluoride content in tobacco and pan masala that are available in the local market of Delhi.[46] They investigated 8 samples of tobacco and 15 samples of pan masala (7 without tobacco and 8 with tobacco). It was found that the fluoride content varied from 28.0 to 113.0 mg/kg for tobacco. In the case of pan masala without tobacco, the fluoride content varied between 23.5 and 185.0 mg/kg, whereas for pan masala with tobacco, it was between 16.5 and 306.5 mg/kg. Fluoride ingestion would be different depending on the intake habits of various people.[46] Kumar and Yadav (2014) estimated the levels of fluoride content in pan masala, chewing tobacco, and betel nuts in the rural and urban areas of Rampur district, Uttar Pradesh, India.[47] Water soluble fluoride content ranged from 23.50 to 42.50 mg/kg in pan masala. In case of chewing tobacco, water-soluble fluoride ranged from 18.10 to 40.00 mg/kg. Water-soluble fluoride in supari (betel nut) was found to be in the range of 8.8–76.50 mg/kg. The intake of two to four sachets of pan masala and chewing tobacco by a person would yield between 0.18 and 1.20 mg and between 0.40 and 1.20 mg of fluoride per day, in addition to the fluoride ingested from food and liquids.[47]

6.4.5 Fluoride from Occupational Exposure

Fluoride is a common pollutant in the industrial workspace. Workers in many heavy industries such as aluminum, fertilizer, iron, oil refining, semiconductor, and steel may be routinely exposed to high levels of fluoride. Fluoride is also used in the welding process; therefore, welders are also commonly exposed to airborne fluorides. As per current U.S. regulations, industries can have fluorides up to 2.5 mg/m^3 in the workspace air, which produces a fluoride intake of 16.8 mg/day for an 8-h working day. A long-term air-monitoring study[48] demonstrated that the fluorine concentration in the air all around the workplace of a small-scale enamel enterprise was in the range of 0.1–3.7 mg/m^3. In the United States, the fluoride concentration in the aluminum production industries for exposed workers was reported to be 1.25 mg/m^3. The average concentration of particulate fluorides was measured as 1.024 mg/m^3, whereas gaseous fluoride (HF) had a mean level[49] of 0.22 mg/m^3. In Sweden, the average TF exposure of the workers in an aluminum plant was calculated as 0.91 mg/m^3, of which 34% (approximately 0.31 mg/m^3) was gaseous fluoride.[50] Electronic industry workers in Japan, where hydrogen fluoride is used for glass etching of TV picture tubes and as a silicon cleaner for semiconductors, are exposed to a daily average concentration of up to 5 mg/L of air hydrogen fluoride.[51] It was reported[52] that the workers in the pot room of an aluminum smelter in British Columbia, Canada, were exposed to an average total airborne fluoride concentration of approximately 0.48 mg/m^3. In Netherlands, the concentrations of fluoride range in the workroom air of welding machine shops and shipyards were reported[2] to be 30–16,500 µg/m^3. From 1981 to 1983, the National Institute for Occupational Safety and Health (NIOSH) conducted the National Occupational Exposure Survey (NOES), which collected data on the effect of occupational exposure of chemical, physical, and biological agents on the workers. The NOES estimated that about 182,589 workers were affected by inhalation of hydrogen fluoride.[53]

As pointed out earlier, large numbers of workers are getting exposed to toxic chemicals while working in industrial areas. Workers in many heavy industries such as aluminum, fertilizer, iron, oil refining, semiconductor, and steel may be routinely exposed to high levels of fluoride. Arshad and Shanavas (2013) studied the effect of fluoride exposure on workers in the fertilizer industry and the wood industry in Mangalore city, India.[54] They investigated the fluoride concentration in the serum and urine of the 34 workers from the fertilizer industry and the 55 workers from the wood industry. Urinary fluoride and serum fluoride levels are valid biomarkers for estimating the levels of occupational exposure to fluoride. The fluoride concentrations in the serum and the urine of the workers employed in the fertilizer industry were 0.077 ± 0.027 and 3.85 ± 1.66 mg/L, respectively. The workers in the wood industry had fluoride concentrations of 0.037 ± 0.009 and 0.97±0.37 mg/L in their serum and urine, respectively. The study concluded that the phosphate fertilizer workers in India are at a high risk of exposure to excessive amounts of fluoride.[54] Sharma et al. (1991) conducted a study to investigate the effect of fluoride in a factory manufacturing inorganic fluoride compounds.[55] The preshift and postshift urinary fluoride levels of workers were estimated. The preshift urinary fluoride levels ranged from 0.5 to 4.54 mg/L and the postshift levels ranged from 0.5 to 13.00 mg/L. The preshift and postshift urinary fluoride concentration depends on the nature of work and the category of workers in each department.[55] Susheela et al. (2013) performed a study on the effect of fluoride exposure on the workers in one of the largest primary aluminum-producing industries located in the north-eastern part of the state of Uttar Pradesh, India.[56] It was observed that smelter workers had a significantly higher fluoride concentration in their urine and serum than non-smelter workers; in addition, the nail fluoride content was higher in smelter workers than in nonsmelter workers. These studies clearly demonstrate that industrial emission of fluoride is a major source of fluoride exposure.[56]

6.4.6 TF Exposure

As discussed in Sections 6.4.1 through 6.4.5, the TF exposure is influenced by different sources and several factors. The factors that affect the fluoride concentration in foodstuffs include fluoride emission sources in the local area, amount of fertilizers and pesticides applied in agricultural activities, and use of fluoridated water in the preparation of food and so forth.[57] The fluoride concentration in the ambient air is influenced by several factors such as nature and type of the industrial sources in the area, the distance from the fluoride sources, the prevailing meteorological conditions and the geological features of the area defined by its topography.[58] The fluoride concentration in water depends on many factors such as the local geological features and proximity to emission sources, mineral constitution of the aquifers, seepage from nearby saline formations, low recharge and dilution rates in the aquifers, peculiarities of the local soil or rock formations, and so forth.[59] However, many scientific studies suggest that the total daily fluoride exposure in a temperate climate when no fluoride is added to the drinking water is approximately equal to 0.6 mg/adult/day; whereas it is around 2 mg/adult/day in a fluoridated area.[60] A range of estimated fluoride intakes as a consequence of exposure to a number of different sources is given in Table 6.2.

6.5 Effects of Fluoride on Laboratory Animals and In Vitro Systems

Considerable research was undertaken to determine the effect of fluoride in laboratory animals. Effects on the skeleton, organs, and tissues have been observed in a variety of studies conducted in rats and rabbits. Both short-term and long-term effects of fluoride exposure for low doses and high doses were investigated by a number of researchers. Some of the studies related to fluoride exposure on laboratory animals are listed in Table 6.3.

TABLE 6.2

Fluoride intake from different sources

Sources of Fluoride Exposure	Age Group of Exposure	Estimated Fluoride Intake, mg/day (mg/kg Body Weight per day)a	Test Insights	Study Area	References
Foodstuffs	Children aged 1–4 years	(0.05)	457 whole-day meals	Poland	Jedra et al.[61]
Toothpaste and diet	1- to 3-year old children	(0.130)	Fluoride intake from diet was measured by the duplicate plate method, and fluoride ingested from dentifrice was determined by subtracting the amount of fluoride recovered after brushing from the amount originally placed onto the child's toothbrush. Samples were carried out by analyzing 33 children.	Brazil	de Almedia et al.[62]
Food group (Grain products, Vegetables, Fruits, Milk products, Meat and poultry , Nuts and seeds, Fats and oils, Sugars and sweets, and Beverages)	3–5 years	0.454	Estimated mean daily fluoride ingestion	Connersville community (Indiana, USA) having water fluoride concentration of 0.16±0.01 mg/L	Jackson et al.[12]
	3–5 years	0.535	Estimated mean daily fluoride ingestion	Richmond community (Indiana, USA), an optimally fluoridated area having water fluoride concentration of 0.90±0.05 mg/L	Jackson et al.[12]
Commercial food for Infant	Infants, 3–8 months	(0.023–0.029)	Estimated mean daily fluoride intake of infants from food	Japan	Tomori et al.[63]
Total diet samples, including drinking water and beverages	Slovenian military personnel	(0.010–0.035)	Range of fluoride intake was calculated by assuming the mean weight of Slovenian military personnel as 70 kg. The amount of fluoride was determined in 20 lyophilized total diet samples obtained from the Slovenian Military.	Slovenia	Ponikvar et al.[64] Vaidya et al.[65]

(Continued)

TABLE 6.2 (*Continued*)

Fluoride intake from different sources

Sources of Fluoride Exposure	Age Group of Exposure	Estimated Fluoride Intake, mg/day (mg/kg Body Weight per day)[a]	Test Insights	Study Area	References
Food Diet, Liquid included water, milk, ready-made beverages and beverages made at home (diluted powder and concentrated fruit juice, etc., and a beverage of tea leaves and wheat ears with tap water).	2–5 years (Moderate fluoride area)	(0.0252)	Estimated the mean concentration of fluoride from the diet ingested by children of two age groups susceptible to dental fluorosis	Two areas of Japan (Moderate fluoride area having mean water fluoride concentration of 0.555 mg/L and relatively low fluoride area having fluoride concentration in the community water in the range of 0.040–0.131 mg/L)	Nohno et al.[66]
	2–5 years (Relatively low fluoride area)	(0.0126)			
	6–8 years (Moderate fluoride area)	(0.0254)			
	6–8 years (Relatively low fluoride area)	(0.0144)			
Solids food, water and other beverages	2–6 years	(0.017)	Mean fluoride intake from food items	Nonfluoridated area of Brazil	Levy et al.[67]
All drinks (Water, Tea, Milk, Soft drink) All foods (Fruit, Vegetable, Soup and gravy, Rice, Bread)	4-year-old children (First area)	0.413	Dietary fluoride intake in children residing in low, medium and high fluoride areas. (The mean fluoride concentrations in drinking water in the three areas were 0.3, 0.6 and 4.0 mg F/L.)	Iran	Zohouri and Rugg-Gunn.[68]
	4-year-old children (Second area)	0.698			
	4-year-old children (Third area)	3.472			
Food (Enjera, homemade bread, kale stew, potato stew, shiro stew, fish stew), beverage (including tea, coffee), and water (used for drinking and cooking).	Adults (Village A)	10.5	Daily dietary fluoride intake by adults from three rural villages of the Ethiopian Rift Valley (Village A uses water with 1.0 mg/L fluoride, village B uses water with 3.0 mg/L fluoride, and village C uses water with 11.5 mg/L fluoride both for food preparation and for drinking)	Ethiopia	Dessalegne et al.[69]
	Adults (Village B)	16.6			
	Adults (Village C)	35.3			

[a] Data in parentheses are the estimated intakes of fluoride, expressed as mg/kg body weight per day, when presented in the reference cited.

TABLE 6.3

Effect of Fluoride Exposure on Laboratory Animals

Laboratory Animal/In Vitro System	Fluoride Dose and Exposure Time	Affected Organ/ System	Toxicological Issues	References
Female Wistar mice	226 mg/L fluoride ion in drinking water from day 15 of pregnancy until day 14 after delivery	Liver	Ingestion of a high amount of fluoride through drinking water may lead to impaired liver function.	Bouaziz et al.[70]
Adult female mice of Swiss Albinos strain	226 mg/L fluoride ion in drinking water from day 15 of pregnancy until day 14 after delivery	Brain	Fluoride intoxication in the early stage of life interfered with brain physiology and induced neurotoxicity in mice.	Bouaziz et al.[71]
Wistar rats	5, 15 or 50 mg/L of fluoride in drinking water over 60 days	Liver and kidney	Exposure fluoride doses (15 and 50 mg/L) caused alterations in the antioxidant system of liver and kidney of rats. However, exposure to 5 mg/L of a fluoride dose causes few changes in the parameters.	Iano et al.[72]
Rats with surgically induced renal deficiency	5, 15 or 50 mg/L of fluoride in drinking water for a period of 6 months	Bones	Fluoridated water of concentrations of 15 and 50 mg/L caused osteomalacia and reduced bone strength in rats, whereas water with a low to moderate fluoride concentration (0 and 5 mg/L) affected neither bone mineralization nor strength in rats.	Turner et al.[73]
Mature female rats	100 and 150 mg/L of fluoride in drinking water for 90 days	Vertebral bone	Fluoridated water causes an increase in bone mass while simultaneously causing a decrease in bone strength/quality, thereby suggesting the negative effect of fluoride on bone quality.	Sogaard et al.[74]

(Continued)

TABLE 6.3 (*Continued*)

Effect of Fluoride Exposure on Laboratory Animals

Laboratory Animal/In Vitro System	Fluoride Dose and Exposure Time	Affected Organ/ System	Toxicological Issues	References
Rats	25 mg/L of fluoride/rat/day for 8 and 16 weeks	Tissue	Drinking of water containing high fluoride may result in tissue damage and other secondary complications.	Shanthakumari et al.[75]
Male Long–Evans rats	0.33 and 0.95 mg/L of fluorine ion in drinking water for 52 weeks	Nervous system	Chronic administration of fluorine in the form of AlF_3 and NaF in the drinking water of rats caused distinct morphological alterations in the brain, including effects on neurons and cerebrovasculature, and may cause injury to the brain.	Varner et al.[76]
Rabbits	Drinking water with fluoride concentrations of 50 and 100 mg/L for 5 months	Blood	Excessive ingestion of fat and fluoride can cause an oxidative stress reaction and increase serum lipid levels either separately or synergistically, which leads to hypercholesterolemia in the experimental rabbits.	Sun et al.[77]
Male Kunming mice	Drinking water with fluoride concentrations of 11, 22, and 45 mg/L and food with 8.40 mg/kg for 180 days	Nervous system	Chronic exposure of fluoride may impair the long-term recognition memory of male mice, enhance the excitement of mice, and upregulate VAMP-2 mRNA expression, all of which are involved in object recognition memory of the nervous system.	Han et al.[78]
Male albino rats	2 mg of sodium fluoride in 1 ml of distilled water per 100 g body weight per day for 29 days	Reproductive system	Fluoride exposure may cause an adverse effect on the reproductive system.	Ghosh et al.[79]

6.6 Effect of Fluoride on Aquatic Organisms

Industrial applications such as phosphate processing, aluminum smelting, steel manufacturing, and glasses frosting are capable of producing an effluent with a high concentration of fluoride. These high fluoride effluents may find their way to the nearby water bodies, thereby causing damage to the aquatic animals and plants. Several researchers gathered information on the effect of fluoride on aquatic organisms. Mishra and Mohapatra (1998) performed a study to measure the fluoride concentration in bones and to monitor the haematological characteristics (RBC, haemoglobin, haematocrit, mean corpuscular haemoglobin, and mean corpuscular volume) in amphibians, *Bufo melanostictus*, collected from fluoride-contaminated and -uncontaminated areas of the Hirakud Smelter Plant, Hirakud, India.[80] The average haemoglobin content, total RBC count and haematocrit in blood samples were significantly reduced, whereas the mean corpuscular concentration and volume were found to significantly increase with respect to the toads at an uncontaminated site. The average fluoride concentration in bones was 2736 mg/kg at the contaminated site, which was 11 times greater than the fluoride concentration in bones of toads from the contaminated areas (241 mg/kg).[80] Hemens and Warwick (1972) performed experiments to determine the short-term and long-term effects of fluoride on fish and prawns in an estuary in Zululand, South Africa.[81] No toxic effects of fluoride were noticed on the species of fish and prawns during their exposure to fluoride up to 100 mg/L for 96 h (short-term exposure). The brown mussel *Pernaperna* showed evidence of toxic effects after the fifth day of fluoride exposure at a concentration of 7.2 mg/L. Long-term (72 days) exposure of fluoride at a concentration of 52 mg/L was performed in recirculated outdoor laboratory estuary models without providing external food and with 20% salinity. Results of long-term exposure showed physical deterioration and an increase in mortality in the mullet *Mugil cephalus* and the crab *Tylodiplax blephariskios*. The reproductive processes of the shrimp *Palaemon pacificus* were found to be adversely affected due to the long-term exposure of fluoride.[81] Johnstone et al. (1982) conducted an experiment to determine the effect of exposure of cryolite recovery sludge (CRS, an aluminum smelter waste dumped at sea) filtrate, which also contains fluoride on salmon fish.[82] The effects of exposure of salmon on CRS filtrate (for up to 1 h) were monitored. It was observed that Atlantic salmon exposed to aluminum smelter waste (including fluoride) experienced an increase in oxygen consumption and ventilation rates and a decrease in heart rate.[82] Shi et al. (2009) carried out an experiment to determine the accumulation of fluoride ion in juvenile sturgeon fish.[83] In a growth trial of 90 days, fishes were exposed to concentrations of 4, 10, 25 and 62.5 mg/L of F (added as NaF), along with a control group. Results indicated that there was a significant inhibition of growth for groups exposed to high fluoride concentrations (10, 25 and 62.5 mg/L) compared with the control group. Shi et al. also observed that exposure of fluoride to a concentration of 25 mg/L or more may cause alterations in the fishes' respiration and violent erratic movements.[83]

6.7 Effect of Fluoride on Plants

Fluoride can enter into plants mainly through two pathways: aerial deposition of gaseous fluoride through stomatal diffusion and passive diffusion through soil and water into the plant roots. Fluoride in the form of gas enters into the stomata of the leaf by diffusion. Initially, it accumulates in the stomata from where it moves toward the tip and the margin, causing injury to the leaf. The injury symptoms are produced only when a critical level of fluoride is attained.[84] Fluoride as particulate falls on the leaf from the polluted atmosphere and gets deposited on the surface of the leaf. Subsequently, these deposited fluorides on the surface penetrate into the leaf and destroy it.[85] The symptoms observed in plants due to the exposure to hydrogen fluoride depend on a number of factors such as the concentration of HF, time of exposure, type and age of plant, temperature, type of light and intensity, composition and rate of circulation of air. When exposed to high concentrations of HF gas for sufficient time under controlled environmental conditions, the plants that are sensitive to the HF may produce one or more of the following effects: slight paling of normal green pigment at the tips or margins of the leaf that may spread to other portions of the leaf; a pale green area at the margins that may gradually turn into a light buff color and finally, a reddish brown. All these stated effects of fluoride influence the photosynthesis and respiration process of plants. The exact mechanism of injury to plants by fluorides is unknown. It has been postulated that they interfere with the functioning of certain enzymes such as enolase.[86] Fluoride can enter into the plant system through the soil; it may also be deposited into the soil from several anthropogenic sources from where it gets accumulated in the plant through roots. The accumulation of fluoride from the soil is generally very small, and there is a limited relationship between concentrations in plants and the total content in soils.[87] Plants absorb fluoride from the soil by their roots, and this gets transported to the transpiratory organs of the plant (mainly the leaves) via xylematic flow. This transported fluoride from the soil can accumulate in the leaves where it can cause adverse effects such as tip burning and even plant death by affecting the photosynthesis and transpiration process.[88,89] A lot of research was undertaken to determine the effect of fluoride on different types of plants and tree species.

Kessabi et al. (1984) performed a study to determine the effect of fluorine emission from the factories processing natural phosphate on plants and animals of south Safi zone (Morocco).[90] These factories were 10 km south of Safi.

Results revealed that the concentrations of fluoride were 4–10 times higher in contaminated plants than in noncontaminated plants. In a certain study area, the effect of fluoride pollution is so high and noticeable that grain crops and vegetables are no longer grown there. In some areas, 30% of the grain corps showed burn signs at the tips of leaves and their yield was reduced. In trees, many fruits either fail or are necrosed. But even in the most contaminated zones, the grasses were unaffected.[90] Zouari et al. (2014) carried out a pot experiment to investigate the uptake, accumulation, and toxicity effects of fluoride in olive trees that were grown in a soil spiked with inorganic fluoride in the form of sodium fluoride.[91] NaF was applied through irrigating water in six different groups of olive with six different concentrations (0, 20, 40, 60, 80 and 100 mM NaF). Symptoms due to fluoride toxicity such as leaf necrosis and leaf drop appeared only in highly spiked soils (80 and 100 mM NaF). It was also reported that a significant reduction of biomass took place in roots, shoots, and leaves of olive plants that are exposed to 60, 80 and 100 mM NaF in comparison to the control plants. But the biomass reduction was not significant for both 20 and 40 mM NaF treated soil.[91] Singh and Verma (2013) performed an experiment to examine the influence of fluoride-contaminated irrigation water having a concentration of fluoride from 100 to 500 mg/L on poplar seedlings (*Populus deltoides* L. clone-S_7C_{15}).[92] Results indicated that the exposure of the poplar seedlings to 100, 200 and 500 mg/L of fluoride in the irrigation water for six weeks decreases the physiological characteristics (growth, leaf expansion, photosynthetic CO_2 assimilation, stomatal conductance, chlorophyll fluorescence yield, and plant biomass). Intervein chlorosis and leaf-margin necrosis followed by leaf curl were observed even in the younger leaves after the exposure of the seedling to irrigation water containing 500 mg/L of fluoride for six weeks. It was also observed that continuous and prolonged exposure of fluoride-contaminated water results in falling of the leaves.[92]

6.8 Effect of Fluoride on Animals

Excessive fluoride injections can affect animals. The impact of fluoride depends on a number of factors such as dosage or amount of intake, rate of intake, period of administration, and the presence of interfering substances. Fluoride effects on animals also depend on the physical parameters of the animal, such as age, state of health, and sensitivity. Young animals are generally more susceptible to harmful effects of fluoride than older ones. Healthy animals have more resistance to the harmful effects of fluoride than sick or inadequately nourished animals.[86] The intake of fluoride in an excessive quantity than required by the animal body can either induce acute toxicosis or cause chronic intoxication depending on the concentration of fluoride and the time of exposure of the animals. The symptom that arises in the animals when they inhale large quantities of fluoride (several grams) in a very short interval of time (few minutes or hours) is termed *acute symptom*. The symptoms that usually arise due to acute toxicosis are an immediate decrease in appetite, high fluoride content in animals' blood and urine, rapid loss of weight, reduced milk production, weakness, excessive salivation, perspiration, dyspnea, and weakened pulse.[93] Animals grazing on fluoride-affected plants may develop fluorosis, characterized by damage to the musculoskeletal system, including difficulty in mastication, softening of the teeth, painful gait, and lameness. Fluorosis occurs in animals grazing in fields near brickworks, aluminum smelters, and phosphate fertilizer factories.

6.9 Guidelines Values and Standards

The WHO guidelines on fluoride in drinking water stipulate that less than 1 mg/L may give rise to dental fluorosis in some children, and much higher concentrations (>1.5 mg/L) may eventually result in skeletal damage in both children and adults. So, in order to prevent dental caries, a large number of communities supply water with a fluoride concentration that is equal to approximately 1.0 mg/L. The 1971 International Standards recommended control limits for fluorides in drinking water for various ranges of the annual average of maximum daily air temperatures. This limit ranges from 0.6 to 0.8 mg/L for a temperature range of 26.3–32.6°C and from 0.9 to 1.7 mg/L for a temperature range of 10–12°C. In the first edition of the *Guidelines for Drinking Water Quality*, published in 1984, a guideline value of 1.5 mg/L was recommended by the WHO for fluoride, as mottling of teeth has been reported very occasionally at higher levels. It was also noted that this guideline value is not fixed and local application must take consideration of local climatic conditions, diet, and water consumption. The 1993 WHO Guidelines concluded that there was no sufficient evidence to suggest that the guideline value of 1.5 mg/L set in 1984 needed to be revised. In some countries, particularly parts of India, Africa, and China, drinking water can contain very high concentrations of naturally occurring fluoride (in excess of the WHO guideline value of 1.5 mg/L). So, it was felt that the guideline value may be difficult to achieve in some circumstances with the treatment technology available. In 1994, the WHO recommended that the optimal concentration of fluoride should be in the range of 0.5–1.0 mg/L and should vary according to climatic conditions, volume of water intake, and intake of fluoride from other sources. Fluoride effects are best predicted by the dose (mg fluoride/kg of body weight/day), the duration of exposure, and other factors such as age. The U.S. National Academy of Sciences Institute of Medicine has recommended an adequate intake of fluoride from all sources as 0.05 mg F/kg body weight/day. This amount exhibits a reduction in the occurrence of dental caries in maximum cases without inducing unwanted side effects, including moderate dental fluorosis.[2,6,94]

6.10 Summary

- Natural processes such as weathering and dissolution of minerals, volcanic eruptions, and marine aerosols are responsible for fluoride release into the environment. Volcanoes are the main natural persistent source of fluorine.
- The transportation, distribution, transformation, and deposition of airborne fluoride are governed by meteorological conditions, particulate size, chemical reactivity, and emission strength of the source.
- The daily fluoride intake of children, expressed in a milligram-per-kilogram body weight basis, may exceed that of adults. It has been found that ingestion of the toothpastes containing fluoride may contribute to 0.50–0.75 mg fluoride per child per day.
- The concentration of the fluoride in the raw food items is the function of fluoride content in the irrigation water.
- Workers in many heavy industries such as aluminum, fertilizer, iron, oil refining, semiconductor, and steel get routinely exposed to high levels of fluoride.
- Urinary fluoride and serum fluoride levels are valid biomarkers for estimating the levels of fluoride due to occupational exposure of fluoride.
- Industrial emission of fluoride is a major source of fluoride exposure.
- Studies on toxicological impacts of fluoride on animals suggest impaired liver function, changes in brain physiology, induced neurotoxicity, alterations in the antioxidant system of liver and kidney, reproductive systems, reduced level of bone quality, tissue damages, morphological alterations and injury to brain, oxidative stress reaction, and so forth.
- Fluoride exposures influence the photosynthesis and respiration process of plants.
- Fluorosis occurs in animals grazing in fields near brickworks, aluminum smelters, and phosphate fertilizer factories.

References

1. Ayoob, S. and Gupta, A.K. (2006). Fluoride in drinking water: A review on the status and stress effects. *Crit. Rev. Environ. Sci. Technol.*, 36, 433–487.
2. WHO. (2002). Environmental Health Criteria 227, Fluorides. Geneva, Switzerland: World Health Organization.
3. Barnard, W.R. and Nordstrom, D.K. (1982). Fluoride in precipitation-II. Implications for the geochemical cycling of fluorine. *Atmos. Environ.*, 16, 105–111.
4. ATSDR. (2003). Report on Toxicological Profile For Fluorides, Hydrogen Fluoride and Fluorine. Atlanta, GA: U.S. Department of Health and Human Services, Public Health Service Agency for Toxic Substances and Disease Registry.
5. Cape, J.N., Fowler, D. and Davison, A. (2003). Ecological effects of sulfur dioxide, fluorides, and minor air pollutants: Recent trends and research needs. *Environ. Int.*, 29, 201–211.
6. WHO. (2006). In: Farewell, J., Bailey, K., Chilton, J., Dahi, E., Fewtrell, L. and Magara, Y. (Eds.) Fluoride in Drinking-Water. World Health Organization. London-Seattle: IWA Publishing.
7. Slooff, W., Eevens, H.C., James, J.A. and Rose, J.R.M. (1989). Integrated Criteria Document Fluoride. The Netherlands: National Institute of Public Health and Environment Pollution, Bilthover (Report No. 758474010).
8. Murray J.J. [Ed.] (1986). Appropriate Use of Fluorides for Human Health. Geneva, Switzerland: World Health Organization.
9. Bralić, M., Buljac, M., Prkić, A., Buzuk, M. and Brinić, S. (2015). Determination fluoride in products for oral hygiene using flow-injection (FIA) and continuous analysis (CA) with home-made FISE. *Int. J. Electrochem. Sci.*, 10, 2253–2264.
10. Kumar, R. and Yadav, S.S. (2014). Fluoride content in pan masala, chewing tobacco, betel nuts (supari), toothpaste and tooth-powder items used and consumed in rural and urban parts of Rampur district, Uttar Pradesh, India. *J. Sci. Technol. Manag.*, 2, 38–47.
11. Yadav, A.K., Kaushik, C.P., Haritash, A.K., Singh, B., Raghuvanshi, S.P. and Kansal, A. (2007). Determination of exposure and probable ingestion of fluoride through tea, toothpaste, tobacco and pan masala. *J. Hazard. Mater.*, 142, 77–80.
12. Jackson, R.D., Brizendine, E.J., Kelly, S.A., Hinesley, R., Stookey, G.K. and Dunipace, A.J. (2002). The fluoride content of foods and beverages from negligibly and optimally fluoridated communities. *Community Dent. Oral Epidemiol.*, 30, 382–91.
13. Liu, C., Wyborny, L.E. and Chan, J.T (1995). Fluoride content of dairy milk from supermarket: A possible contributing factor to dental fluorosis. *Int. Soc. Fluoride Res.*, 28, 10–16.
14. Amalraj, A. and Pius, A. (2013). Health risk from fluoride exposure of a population in selected areas of Tamil Nadu South India. *Food Sci. Hum. Wellness*, 2, 75–86.
15. Duff, E.J. (1981). Total and ionic fluoride in milk. *Caries Res.*, 15, 406–408.
16. Dabeka, R.W. and McKenzie, A.D. (1987). Lead, cadmium, and fluoride levels in market milk and infant formulas in Canada. *J. Assoc. Off. Anal. Chem.*, 70, 754–757.

17. Buzalaf, M.A., de Almeida, B.S., Cardoso, V.E., Olympio, K.P. and Furlani, T. de A. (2004). Total and acid-soluble fluoride content of infant cereals, beverages and biscuits from Brazil. *Food Addit. Contam.*, 21, 210–5.
18. Fein, N.J. and Cerklewski, F.L. (2001). Fluoride content of foods made with mechanically separated chicken. *J. Agric. Food Chem.*, 49, 4284–4286.
19. Jedra, M., Urbanek-Karłowska, B., Fonberg-Broczek, M., Sawilska-Rautenstrauch, D. and Badowski, P. (2001). Bioavailable fluoride in poultry deboned meat and meat products. *Rocz. Panstw. Zakl. Hig.*, 52, 225–230.
20. Heilman, J.R., Kiritsy, M.G., Levy, S.M. and Wefel, J.S. (1997). Fluoride concentrations of infant foods. *J. Am. Dent. Assoc.*, 128, 857–863.
21. Camargo, J. (2003). Fluoride toxicity to aquatic organisms: A review. *Chemosphere*, 50, 251–264.
22. Rocha, R.A., Rojas, D., Ruiz, A., Devesa, V. and Ve, D. (2013). Quantification of Fluoride in Food by Microwave Acid Digestion and Fluoride Ion-Selective Electrode. *J. Agric. Food Chem.*, 16, 10708–10713.
23. Thomas, A. and James, R. (2013). Accumulation of Fluoride in *Etroplus Suratensis, Oreochromis Mossambicus and Anabas Testudineus* Caught from the Surface Fresh Water Sources in Alappuzha town, Kerala, India. *Int. J. Innov. Res. Sci. Eng. Technol.*, 2, 2756–2761.
24. Sawilska-Rautenstrauch, D., Jedra, M., Fonberg-Broczek, M., Badowski, P. and Urbanek-Karłowska, B. (1998). Fluorine in vegetables and potatoes from the market in Warsaw. *Rocz. Panstw. Zakl. Hig.*, 49, 341–346.
25. Thippeswamy, H.M., Kumar, N., Anand, S.R., Prashant, G.M. and Chandu, G.N. (2010). Fluoride content in bottled drinking waters, carbonated soft drinks and fruit juices in Davangere city, India. *Indian J. Dent. Res.*, 21, 528–530.
26. Jimenez-Farfan, M.D., Hernandez-Guerrero, J.C., Loyola-Rodriguez, J.P. and Ledesma-Montes, C. (2004). Fluoride content in bottled waters, juices and carbonated soft drinks in Mexico City, Mexico. *Int. J. Paediatr. Dent.*,14, 260–266.
27. Heilman, J.R., Kiritsy, M.C., Levy, S.M. and Wefel, J.S. (1999). Assessing fluoride levels of carbonated soft drinks. *J. Am. Dent. Assoc.*, 130, 1593–1599.
28. Lung, S.C., Cheng, H.W. and Fu, C.B. (2007). Potential exposure and risk of fluoride intakes from tea drinks produced in Taiwan. *J. Expo. Sci. Environ. Epidemiol.*, 18, 158–166.
29. Wong, M., Fung, K. and Carr, H. (2003). Aluminium and fluoride contents of tea, with emphasis on brick tea and their health implications. *Toxicol. Lett.*, 137, 111–120.
30. Opydo-szymaczek, J. and Opydo, J. (2009). Fluoride content of bottled waters recommended for infants and children in Poland. *Res. Rep. Fluoride*, 42, 233–236.
31. Aldrees, A.M. and Al-Manea, S.M. (2010). Fluoride content of bottled drinking waters available in Riyadh, Saudi Arabia. *Saudi Dent. J.*, 22, 189–93.
32. Bengharez, Z., Farch, S., Bendahmane, M., Merine, H. and Benyahia, M. (2012). Evaluation of fluoride bottled water and its incidence in fluoride endemic and non endemic areas. *e-SPEN. J.*, 7, e41–e45.
33. Rao, K.V. and Mahajan, C.L. (1990). Fluoride content of some common South Indian foods and their contribution to fluorosis. *J. Sci. Food. Agric.*, 51, 215–219.
34. Gautam, R., Bhardwaj, N. and Saini, Y. (2010). Fluoride accumulation by vegetables and crops grown in Nawa Tehsil of Nagaur district (Rajasthan, India). *J. Phytol.*, 2, 80–85.

35. Bhargava, D. and Bhardwaj, N. (2009). Study of fluoride contribution through water and food to human population in fluorosis endemic villages of North-Eastern Rajasthan. *Afr. J. Basic Appl. Sci.*, 1, 55–58.
36. Raghavachari, S., Tripathi, R.C. and Bhupathi R.K. (2008). Endemic fluorosis in five villages of the Palamau district, Jharkhand, India. *Fluoride*, 41, 206–211.
37. Ramteke, D. S., Onkar, R., Pakhide, D. and Sahasrabudhe, S. (2007). Assessment of Fluoride in Groundwater, Food and Soil and its Association with Risk to Health. Proceedings of the 10th International Conference on Environmental Science and Technology, Kos Island, Greece.
38. Yadav, R.K., Sharma, S., Bansal, M., Singh, A., Panday, V. and Maheshwari, R. (2012). Effects of fluoride accumulation on growth of vegetables and crops in Dausa district, Rajasthan, India. *Adv. Biores.*, 3, 14–16.
39. Kabata Pendias, A. and Pendias, H. (2001). *Trace Elements in Soils and Plants*, 3rd edn, p. 413. Boca Raton, FL: CRC Press.
40. Cronin, S.J., Manohara, V., Hedley, M.J. and Loganathan, P. (2000). Fluoride: A review of its fate, bioavailability, and risks of fluorosis in grazed-pasture systems in New Zealand. *N. Z. J. Agri. Res.*, 43, 295–321.
41. Jha, S.K. (2012). Geochemical and spatial appraisal of fluoride in the soils of indo-gangetic plains of India using multivariate analysis. *Clean Soil Air Water*, 40, 1392–1400.
42. Mishra, P.C., Meher, K., Bhosagar, D. and Pradhan, K. (2009). Fluoride distribution in different environmental segments at Hirakud Orissa (India). *Sci. Technol.*, 3, 260–264.
43. Jha, S.K., Nayak, A. K., Sharma, Y.K., Mishra, V.K. and Sharma, D.K. (2008). Fluoride accumulation in soil and vegetation in the vicinity of brick fields. *Bull. Environ. Contam. Toxicol.*, 80, 369–373.
44. Chaudhary, V., Sharma, M. and Yadav, B.S. (2009). Elevated fluoride in canal catchment soils of Northwest Rajasthan, India. *Res. Rep. Fluoride*, 42, 46–49.
45. Anbuvel, D., Kumaresan, S. and Margret, R.J. (2014). Flouride analysis of soil in cultivated areas of Thovalai channel in Kanyakumari District, Tamilnadu, India: Correlation with physico-chemical parameters. *Int. J. Basic Appl. Chem. Sci.*, 4, 20–29.
46. Yadav, A.K., Kaushik, C.P., Haritash, A.K., Singh, B., Raghuvanshi, S.P. and Kansal, A. (2007). Determination of exposure and probable ingestion of fluoride through tea, toothpaste, tobacco and pan masala. *J. Hazard. Mater.*, 142, 77–80.
47. Kumar, R. and Yadav, S.S. (2014). Fluoride content in pan masala, chewing tobacco, betel nuts (supari), toothpaste and tooth-powder items used and consumed in rural and urban parts of Rampur District, Uttar Pradesh, India. *J. Sci. Technol. Manag.*, 2, 38–47.
48. Viragh, E., Viragh, H., Laczka, J. and Coldea, V. (2006). Health effects of occupational exposure to fluorine and its compounds in a small-scale enterprise. *Ind. Health*, 44, 64–68.
49. Taiwo, O.A., Sircar, K.D., Slade, M.D., Cantley, L.F., Vegso, S.J., Rabinowitz, P.M., Fiellin, M.G. and Cullen, M.R. (2006). Incidence of asthma among aluminium workers. *J. Occup. Environ. Med.*, 48, 275–282.
50. Ehrnebo, M. and Ekstrand, J. (1986). Occupational fluoride exposure and plasma fluoride levels in man. *Int. Arch. Occup. Environ. Health*, 58, 179–190.

51. Kono, K., Yoshida, Y. and Yamagata, H. (1987). Urinary fluoride monitoring of industrial hydrofluoric acid exposure. *Environ. Res.*, 42, 415–520.
52. Chan-Yeung, M., Wong, R., Earnson, D., Schulzer, M., Subbarao, K., Knickerbocker, J. and Grzybowski, S. (1983). Epidemiological health study of workers in an aluminum smelter in Kitimat, B.C. II. Effects on musculoskeletal and other systems. *Arch. Environ. Health*, 38, 34–40.
53. NIOSH. (1989). National occupational exposure survey (1980–1983). Cincinnati, OH: National Institute for Occupational Safety and Health, Department of Health and Human Services.
54. Arshad, M. and Shanavas, P. (2013). Comparison of serum and urinary fluoride levels among fertilizer and wood industry workers in Mangalore city, India. *Res. Rep. Fluoride*, 46, 80–82.
55. Sharma, Y.K., Kulkarni, P.K., Shah, A.R., Patel, M.D. and Kashyap, S.K. (1991). Occupational exposure to inorganic fluorides. *Indian J. Ind. Med.*, 37, 13–22.
56. Susheela, A.K., Mondal, N.K. and Singh, A. (2013). Exposure to fluoride in smelter workers in a primary aluminum industry in India. *Int. J. Occup. Environ. Med.*, 4, 61–72.
57. Myers, H.M. (1978). Fluorides and dental fluorosis. *Monogr. Oral. Sci.*, 7, 1–76.
58. Davis, W.L. (1972). Ambient air fluorides in Salt Lake County, Rocky Mountain. *Med. J.*, 69, 53–56.
59. Hudak, P.F. (1999). Fluoride levels in Texas groundwater. *J. Environ. Sci. Heal. A*, 34, 1659–1676.
60. WHO. (1984). Fluorine and Fluorides, Environmental Health Criteria 36. Geneva, Switzerland: World Health Organization.
61. Jędra, M., Sawilska-rautenstrauch, D., Gawarska, H. and Starski, A. (2011). Fluorine content in total diets samples of small children. *Rocz. Panstw. Zakl. Hig.*, 62, 275–281.
62. de Almeida, B.S., da Silva Cardoso, V.E. and Buzalaf, M.A. (2007). Fluoride ingestion from toothpaste and diet in 1- to 3-year-old Brazilian children community. *Dent. Oral Epidemiol.*, 35, 53–63.
63. Tomori, T., Koga, H., Maki, Y. and Takaesu, Y. (2004). Fluoride analysis of foods for infants and estimation of daily fluoride intake. *Bull. Tokyo Dent. Coll.*, 45, 19–32.
64. Ponikvar, M., Stibilj, V. and Žemva, B. (2007). Daily dietary intake of fluoride by Slovenian military based on analysis of total fluorine in total diet samples using fluoride ion selective electrode. *Food Chem.*, 103, 369–374.
65. Vaidya, R., Bhalwar, R. and Bobdey, S. (2009). Anthropometric parameters of armed forces personnel. *Med. J. Armed Forces India*, 65, 313–318.
66. Nohno, K., Sakuma, S., Koga, H., Nishimuta, M., Yagi, M. and Miyazaki, H. (2006). Fluoride intake from food and liquid in Japanese children living in two areas with different fluoride concentrations in the water supply. *Caries Res.*, 40, 487–493.
67. Levy, F.M., Olympio, K.P., Philippi, S.T. and Buzalaf, M.A. (2013). Fluoride intake from food items in 2- to 6-year-old Brazilian children living in a non-fluoridated area using a semiquantitative food frequency questionnaire. *Int. J. Paediatr. Dent.*, 23, 444–451.
68. Zohouri, F.V. and Rugg-Gunn, A.J. (2000). Sources of dietary fluoride intake in 4-year-old children residing in low, medium and high fluoride areas in Iran. *Int. J. Food Sci. Nutr.*, 51, 317–26.

69. Dessalegne, M. and Zewge, F., (2013). Daily dietary fluoride intake in rural villages of the Ethiopian Rift Valley. *Toxicol. Environ. Chem.*, 95, 1056–1068.
70. Bouaziz, H., Ketata, S., Jammoussi, K., Boudawara, T., Ayedi, F., Ellouze, F. and Zeghal, N. (2006). Effects of sodium fluoride on hepatic toxicity in adult mice and their suckling pups. *Pestic. Biochem. Physiol.*, 86, 124–130.
71. Bouaziz, H., Amara, I. Ben, Essefi, M., Croute, F. and Zeghal, N. (2010). Fluoride-induced brain damages in suckling mice. *Pestic. Biochem. Physiol.*, 96, 24–29.
72. Iano, F.G., Ferreira, M.C., Quaggio, G.B., Fernandes, M.S., Oliveira, R.C., Ximenes, V.F. and Buzalaf, M.A.R. (2014). Effects of chronic fluoride intake on the antioxidant systems of the liver and kidney in rats. *J. Fluor. Chem.*, 168, 212–217.
73. Turner, C.H., Owan, I., Brizendine, E.J., Zhang, W., Wilson, M.E. and Dunipace, A. J. (1996). High fluoride intakes cause osteomalacia and diminished bone strength in rats with renal deficiency. *Bone*, 19, 595–601.
74. Sogaard, C.H., Mosekilde, L., Schwartz, W., Leidig, G., Minne, H.W. and Ziegler, R. (1995). Effects of fluoride on rat vertebral body biomechanical competence and bone mass. *Bone*, 16, 163–169.
75. Shanthakumari, D., Srinivasalu, S. and Subramanian, S. (2004). Effect of fluoride intoxication on lipidperoxidation and antioxidant status in experimental rats. *Toxicology*, 204, 219–228.
76. Varner, J., Jensen, K.F., Horvath, W. and Isaacson, R.L. (1998). Chronic administration of aluminum-fluoride or sodium-fluoride to rats in drinking water: Alterations in neuronal and cerebrovascular integrity. *Brain Res.*, 784, 284–298.
77. Sun, L., Gao, Y., Zhang, W., Liu, H. and Sun, D. (2014). Effect of high fluoride and high fat on serum lipid levels and oxidative stress in rabbits. *Environ. Toxicol. Pharmacol.*, 38, 1000–1006.
78. Han, H., Du, W., Zhou, B., Zhang, W., Xu, G., Niu, R. and Sun, Z. (2014). Effects of chronic fluoride exposure on object recognition memory and mRNA expression of SNARE complex in hippocampus of male mice. *Biol. Trace Elem. Res.*, 158, 58–64.
79. Ghosh, D., Das, S., Maiti, R., Jana, D. and Das, U.B. (2002). Testicular toxicity in sodium fluoride treated rats: Association with oxidative stress. *Reprod. Toxicol.*, 16, 385–390.
80. Mishra, P.C. and Mohapatra, K. (1998). Haematological characteristics and bone fluoride content in *Bufo melanostictus* from an aluminium industrial site. *Environ. Pollut.*, 99, 421–423.
81. Hemens, J. and Warwick, R.J. (1972). The effects of fluoride on estuarine organisms. *War. Res.*, 6, 1301–1308.
82. Johnstone, A.D.F. and Hawkins, A.D. (1982). The effects of an industrial waste (cryolite recovery sludge) upon the Atlantic salmon, *Salmo salar* (L). *Water Res.*, 16, 1529–1535.
83. Shi, X., Zhuang, P., Zhang, L., Feng, G., Chen, L., Liu, J., Qu, L. and Wang, R. (2009). The bioaccumulation of fluoride ion (F^-) in Siberian sturgeon (*Acipenser baerii*) under laboratory conditions. *Chemosphere*, 75, 376–380.
84. Sharma, M.R. and Gupta, V. (2014). Fluoride and its ecological effects in water: A review. *Global J. Res. Anal.*, 3, 2277–8160.
85. Bellomo, S., Aiuppa, A., D'Alessandro, W. and Parello, F. (2007). Environmental impact of magmatic fluorine emission in the Mt. Etna area. *J. Volcanol. Geotherm. Res.*, 165, 87–101.

86. Greenwood, D.A. (1956). Some Effects of Inorganic Fluoride on Plants, Animals, and Man, *USU Faculty Honor Lectures.* Paper 41. http://digitalcommons.usu.edu/honor_lectures/41.
87. Weinstein, L.H. (1977). Fluoride and plant life. *J. Occup. Med.*, 19, 49–78.
88. Klumpp, A., Klumpp, G., Domingos, M. and Silva, M.D. (1996). Fluoride impact on native tree species of the Atlantic Forest near Cubatao, Brazil. *Water Air Soil Poll.*, 78, 57–71.
89. Davison, A. and Weinstein, L.W. (1998). The effects of fluorides on plants. *Earth Island J.*, 13, 257–264.
90. Kessabi, M., Assimi, B. and Braun, J.P. (1984). The effects of fluoride on animals and plants in the south Safi zone. *Sci. Total Environ.*, 38, 63–68.
91. Zouari, M., Ben Ahmed, C., Fourati, R., Delmail, D., Ben Rouina, B., Labrousse, P. and Ben Abdallah, F. (2014). Soil fluoride spiking effects on olive trees (*Olea europaea* L. cv. Chemlali). *Ecotoxicol. Environ. Saf.*, 108, 78–83.
92. Singh, M. and Verma, K.K. (2013). Influence of fluoride-contaminated irrigation water on physiological responses of poplar seedlings (*Populus deltoides* L. clone-S7C15). *Fluoride*, 46, 83–89.
93. Hobbs, C.S., Moorman, R.P, Griffith, J.M., West, J.L., Merriman, G.M., Hansard, S.L., Chamberlain, C.C., MacIntire, W.M., Hardin, L.J. and Jones, L.S. (1954). Fluorosis in cattle and sheep. *Tenn. Agr. Exp. Sta. Bull.*, 235.
94. WHO. (2003). Fluoride in Drinking-Water. Background document for preparation of WHO Guidelines for drinking-water quality. Geneva, Switzerland: World Health Organization.

7

Defluoridation Techniques: An Overview

7.1 Introduction

The defluoridation techniques generally practiced include (1) coagulation, (2) adsorption (including ion exchange), (3) electrochemical methods, and (4) membrane processes. Coagulation processes mainly use chemical reagents such as lime, calcium, or magnesium salts, poly aluminum chloride, and alum to make precipitation (or co precipitation) with fluoride, necessitating its removal. Adsorption is a popular technique practiced in fluoride endemic areas of the developing world. In this method, the adsorbent is used in fixed columns in packed beds and fluoride-laced water is cycled through it. The pollutant from a relatively bulk liquid volume gets concentrated and confined onto a small adsorbent mass, which can invariably be regenerated, reused, or safely disposed under control.[1,2] Electrochemical techniques mainly include electrocoagulation and other electrosorptive processes.[3,4] Electrosorptive techniques basically involve activation of an adsorbent bed and enhanced removal by application on an electric field. Electrocoagulation involves the use of aluminum electrodes that release Al^{3+} ions (by an anodic reaction) that react with fluoride ions near the anode. In this process, the removal of fluoride by precipitation is expected to occur at the electrode–electrolyte interface. Membrane techniques generally include reverse osmosis (RO), nanofiltration (NF), ultrafiltration (UF), electrodialysis, and Donnan dialysis. A combination of two or more of these membrane techniques for enhanced removal of fluoride was also reported.[1,5,6]

7.2 Coagulation

In general, the removal mechanisms that are operative in coagulation include: (1) charge neutralization of negatively charged colloids by cationic hydrolysis products and (2) incorporation of impurities onto an amorphous precipitate of metal hydroxide. The relative importance of these two mechanisms depends on many factors, which are mainly pH and coagulant dosage.[7] Defluoridation processes by coagulation include (1) precipitation of fluoride by a suitable reagent through chemical reactions; (2) co precipitation of fluoride, which involves its simultaneous precipitation with a macro-component from the same solution through the formation of mixed crystals, by multiple mechanisms such as adsorption, occlusion, or mechanical entrapment.[1]

7.2.1 Lime

Precipitation of fluoride in the form of insoluble calcium fluoride (CaF_2) is one of the most commonly adopted precipitation techniques used in defluoridation. For this purpose, either lime [$Ca(OH)_2$] or salts of calcium such as $CaSO_4$ or $CaCl_2$ may be used. The precipitation reaction involves the following:

$$Ca(OH)_2 + 2\,F^- \rightarrow CaF_2 \downarrow + 2OH^- \quad (7.1)$$

If $Ca(OH)_2$ is used as a source of lime, the pH will increase with calcium dosage as per Equation 7.1, which displays the major limitation of the process. Hence, proper dosage of lime should be ensured to keep pH within permissible limits prescribed for drinking water.[8] It has been reported that liming usually leaves higher residual fluoride concentrations of 10– 20 mg/L, which makes drinking water unpalatable.[1,9] So, additional defluoridation processes are to be employed for removing the excess fluoride present. This further demands additional necessities to be provided, such as usage of more chemicals, reagents, and employing processes that enhance the treatment expenses. However, these techniques may render large volumes of additional sludge.[10,11] Another noticeable limitation with lime precipitation is the poor settling characteristics of the precipitate. As stated earlier, "water treated with lime frequently has much higher concentrations of residual fluoride because of slow nucleation during precipitation, leading to a high ionic strength" and hardness.[1,10,12] The supplementary processes to be employed in removing excess chemicals lead to the inappropriateness of this method in ensuring the quality and originality of natural waters.[11] Further, this may add to the total process cost. These constraints that are related to precipitation, limit its usage as a sustainable option for defluoridation of drinking water.

7.2.2 Magnesium Oxide

Magnesium oxide has also been used for defluoridation of drinking water.[13] On hydration, magnesium oxide gets converted to magnesium hydroxide, which combines with fluoride ions and forms practically insoluble magnesium fluoride as follows:

$$MgO + H_2O \rightarrow Mg(OH)_2 \quad (7.2)$$

$$Mg(OH)_2 + 2F^- \rightarrow MgF_2 \downarrow + 2OH^- \tag{7.3}$$

The use of magnesium oxide is prevalent in fluoride endemic areas[14] in domestic defluoridation units (DDUs). However, as shown in Equation 7.3, there will be an increase in the pH of treated water to the range of 10 to 11. However, the addition of small amounts (0.15–0.20 g/L) of sodium bisulfate can decrease pH within the desirable limits (6.5–8.5). The DDU consists of two units of 20 L capacity each. The mixing of magnesium oxide is carried out in the upper unit through a manually operated, geared mechanical stirring device. The lower unit serves as a collection unit. The mixing of the coagulant is completed within 5 min, and the mixture is clarified for 16 h. The bottom portion of the container receives the flocs or sludge that settles in the system. Clear water is poured into the lower collection unit through an elastic connecting pipe. This pipe is fitted with a fine filter for arresting the flocs and tiny sludge elements. In the lower unit, sodium bisulfate is added so as to dissolve it in drinking water before it is supplied for drinking.[1]

7.2.3 Calcium and Phosphate Compounds

Calcium and phosphate compounds such as calcium chloride ($CaCl_2 \cdot 2H_2O$) and monosodium phosphate ($NaH_2PO_4 \cdot H_2O$) can be used for defluoridation in which fluoride gets precipitated as calcium fluoride or fluorapatite, as follows:[1]

$$CaCl_2 \cdot 2H_2O = Ca^{2+} + 2Cl^- + 2H_2O \tag{7.4}$$

$$NaH_2PO_4 \cdot H_2O = PO_4^{3-} + Na^+ + 2H^+ + H_2O \tag{7.5}$$

$$Ca^{2+} + 2F^- = CaF_2 \tag{7.6}$$

$$10Ca^{2+} + 6PO_4^{3-} + 2F^- = Ca_{10}(PO_4)_6F_2 \tag{7.7}$$

This process can be catalyzed in a contact bed consisting of a saturated bone charcoal medium that may act as a filter for the precipitate.[15] In addition, gravel or coarse-grained bone charcoal can be used as a supporting medium for this adsorbent bed in a column. Fluoride-rich water mixed with calcium chloride and monosodium phosphate is fed to the column for a contact period of 20–30 min. The treated water flows continuously by gravity through the bed to the clean water collecting tank. This process has been practiced in many African countries such as Tanzania and Kenya for treating groundwater with fluoride concentrations of about 10 mg/L.[1,16]

7.3 Co Precipitation of Fluoride

7.3.1 Alum

Alum $[Al_2(SO_4)_3 \cdot 18H_2O]$ is a popular coagulant used for the flocculation of colloidal impurities in water. When aluminum (Al) salts are dissolved in water, the metal ion Al^{3+} gets hydrated and forms an aquometalion $Al(H_2O)_6^{3+}$, which upon further hydrolysis forms a series of mononuclear, dinuclear, and possibly polynuclear hydroxo complexes, namely, $Al_{13}(OH)_{34}^{5+}$, $Al_7(OH)_{17}^{4+}$, $Al_8(OH)_{20}^{4+}$ and $Al_6(OH)_{15}^{3+}$ ultimately precipitating onto the metal hydroxide floc of $Al(OH)_3$. The hydrolytic reactions can be expressed as follows:[1,17–19]

$$Al^{3+} + H_2O \leftrightarrow Al(OH)^{2+} + H^+ \tag{7.8}$$

$$Al^{3+} + 2H_2O \leftrightarrow Al(OH)_2^+ + 2H^+ \tag{7.9}$$

$$7Al^{3+} + 17H_2O \leftrightarrow Al_7(OH)_{17}^{4+} + 17H^+ \tag{7.10}$$

$$Al(H_2O)_6^{3+} + H_2O \leftrightarrow Al(H_2O)_5(OH)_2 + H_3O^+ \tag{7.11}$$

$$Al^{3+} + 3H_2O \leftrightarrow Al(OH)_3(s) + 3H^+ \tag{7.12}$$

$$Al^{3+} + 4H_2O \leftrightarrow Al(OH)_4^- + 4H^+ \tag{7.13}$$

Except the simple aquometal ions, the hydroxometal complexes thus formed get readily adsorbed at interfaces[19] that are responsible for the destabilization of colloids in water that are treated with aluminum salts through charge neutralization. Also, "polymers of high molecular weight can adsorb simultaneously on two or more particles and bind them" together through "polymer bridging."[1,7] The acidity of treated water may increase due to the release of H^+, as shown in the earlier equations. So, the pH of the system after alum treatment may be influenced by the dosage of alum and the initial alkalinity of the water treated. In this process, it is plausible that fluoride ions are removed by forming a part of the gelatinous $Al(OH)_3$ flocs, which, subsequently, gets precipitated. It is suggested that the mechanism of fluoride removal due to alum addition may be due to (1) coprecipitation of fluoride and hydroxide (OH) ions with aluminum (Al) ions, forming a precipitate or floc with the chemical formula $Al_nF_m(OH)_{3n-m}$ (Equation 7.14); (2) and/or by adsorption or ligand exchange (Equation 7.15). The fluoride removal mechanism by adsorption or complexation with $Al(OH)_3(s)$ can be expressed as follows:[10,20]

$$nAl_{(aq)}^{3+} + (3n\text{-}m)\,OH_{(aq)}^{-} + mF_{(aq)}^{-} \rightarrow Al_nF_m(OH)_{3n\text{-}m\,(s)} \qquad (7.14)$$

$$Al_n(OH)_{3n\,(s)} + m\,F_{(aq)}^{-} \rightarrow Al_nF_m(OH)_{3n\text{-}m\,(s)} + m\,OH_{(aq)}^{-} \qquad (7.15)$$

Though the species $Al(OH)_4^-$ aids fluoride removal through a ligand-exchange mechanism (Equation 7.16), it may release high aluminum residuals into the treated water.[17]

$$Al(OH)_4^- + F^- \rightarrow Al(OH)_3F + OH^- \qquad (7.16)$$

The "efficiency of removal of fluoride by a fixed alum dose depends on pH, alkalinity, the coexisting anions, and other characteristics of the solution."[1,21] Literature also suggests some major limitations of alum treatment[22,23] due to the increased release of sulfate and aluminum concentrations into treated water. Coagulation with alum results in aluminum residuals of 0.37 mg/L at a pH of 9.8. However, the amount of aluminum residuals released was only 0.07 mg/L at a lower pH of 7.61. This clearly demonstrates that the lowering of pH dramatically reduces the residual aluminum.[17] So, the most appropriate pH for defluoridation by coagulation is in the range of 5.5–6.5.[24] Incidentally, for practical applications, post-treatment pH control may be necessary for providing stable water, which may invite increased initial investments, capital, operational, and maintenance costs, and enhanced hardness of treated water. So, alum treatment is confined to high dosage requirement, issues of sludge disposal, high pH of the treated water, and residual alumina in treated water.[1,25]

7.3.2 Alum and Lime (Nalgonda Technique)

This technique derived its name from Nalgonda, which is a place in India where the first community defluoridation plant was constructed and is a district affected by severe fluorosis. Defined quantities of alum, lime, and bleaching powder are added to raw water; this is followed by rapid mixing, flocculation, sedimentation, filtration, and disinfection.[26] The added sodium aluminate or lime hastens settlement of precipitate, and bleaching powder ensures disinfection. The dose of lime is only 1/20th to 1/25th that of filter alum. Bleaching powder is added to raw water at the rate of 3 mg/L to ensure disinfection. The process involves coagulation with alum in an alkaline aqueous environment; this is followed by adsorption and charge neutralization. "Fluoride may be adsorbed onto the sticky gelatinous $Al(OH)_3$ flocs during sweep coagulation" and may get co precipitated. The use of aluminum sulfate and lime for defluoridation has its genesis from the United States in the 1930s and got popularized in India as the Nalgonda technique after the 1970s. Over the years, in many developing countries such as Tanzania, Senegal, Kenya, and India, this technique has been successfully implemented for community applications and at individual household levels.[5,15,16] "Fill and draw type" defluoridation units are basically designed for community applications for serving around 200 people. The entire operations of such a unit can be completed within 2–3 h with multiple batch performance in a single day. Of late, an advanced electrically operated model of this unit has also been developed.[27] Community installations using "fill and draw type" units having capacities up to 20–40 million gallons have been used in the fluoride endemic areas in India (Figure 7.1). In addition to defluoridation, the mechanisms involved, namely, coagulation, flocculation, clarification, and disinfection, also aid in the simultaneous removal of color, odor, turbidity, bacteria, and organic contaminants from water. The low cost and ease of handling made this process more preferable.[28] However, it is also reported that this technology had limited field applications both as hand pump–based units and as smaller domestic units, mainly due to the need for constant attention. Limitations of this technique also include medium efficiency, high aluminum sulfate dosage, controlling of varying alum and lime dosages for different sources of raw water with different alkalinity and fluoride concentrations, residual sulfate, salinity, hardness of the treated water, higher pHs, and high residual aluminum concentrations.[1,5,16,29]

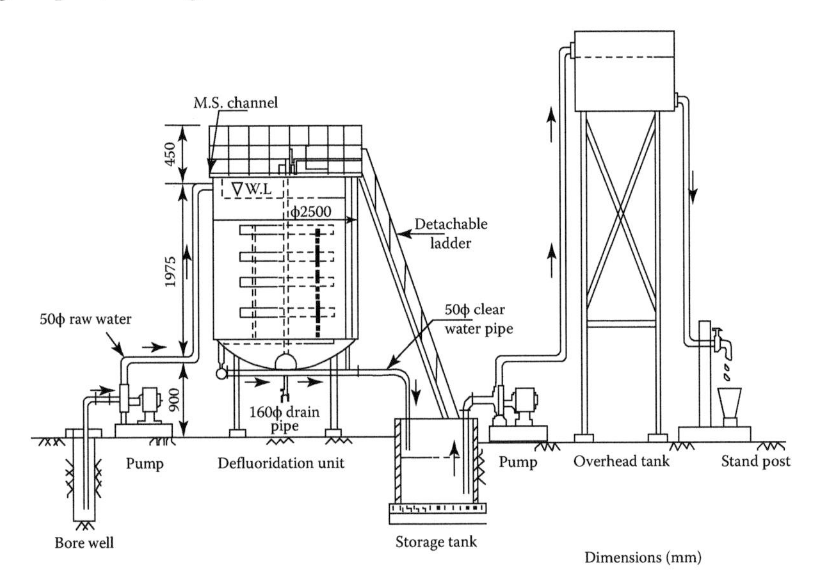

FIGURE 7.1
The fill-and-draw type of defluoridation system for rural water supply for a population of more than 1500 at 40 lpcd. (From CPHEEO, *Central Public Health and Environmental Engineering Organization, Manual on Water Supply and Treatment*, 3rd edn., pp. 289–297, The Controller of Publications, New Delhi, India, 1991.)

7.4 Adsorption

Adsorption basically denotes an interface accumulation of substances at a surface or interface. The material adsorbed is termed *adsorbate,* and the adsorbing phase is called *adsorbent.* The mechanisms of adsorption may be "physisorption," or "chemisorption," or both. A physisorbed species is not attached to a specific site; rather, it is free of any translational motion within the interface. Physisorption may be significant at low temperatures and it develops a low energy of adsorption, indicating that the adsorbate is loosely held with the adsorbent. If the adsorbate develops a chemical interaction with the adsorbent, the process may be referred to as *chemisorption.* The adsorbed molecules are attached on the surface as they form strong localized bonds at the active centers of the adsorbent. The "ion exchange" may be treated as an "exchange adsorption," in which "ions of one substance concentrate at a surface as a result of electrostatic attraction to charged sites at the surface."[1,19] Of late, adsorption or ion exchange is one of the most frequently used methods for defluoridation. Water laced with fluoride is passed through a column packed with an adsorbent and on saturation, the adsorbent bed is backwashed for reuse. The adsorption capacity, cost of the adsorbent, ease in operation, and potential for reuse and regeneration are some of the factors that define the selection of an adsorbent. It would be interesting to understand the mechanism of fluoride removal in the applications of some of the most frequently used adsorbents.

7.4.1 Bone and Bone Charcoal

The use of bone in fluoride scavenging was demonstrated from 1930 onward. The caustic-and-acid–treated bone material was demonstrated successful in reducing fluoride concentration from 3.5 mg/L to even less than 0.2 mg/L. The removal mechanism includes an ion exchange of carbonate radical of the apatite [$Ca_9(PO_4)_6 \cdot CaCO_3$] in the bone with fluoride (forming insoluble fluorapatite) as follows:

$$Ca_9(PO_4)_6 \cdot CaCO_3 + 2F^- \rightarrow Ca_9(PO_4)_6 \cdot CaF_2 + CO_3^{2-} \qquad (7.17)$$

Although the high cost of bone was an inhibiting factor in the initial period, it was reported that bone char (obtained by carbonizing bone at 1100°C–1600°C) has superior defluoridation potential than the original unprocessed bone. As a result, thereafter, bone char was used in defluoridation operations. Bone char is obtained by charring animal bones for removing all organics. The resultant product essentially consists of tricalcium phosphate and carbon. The adsorption mechanism of bone char is also an ion exchange in which phosphate in bone char is exchanged with a fluoride ion.[30] The removal of fluoride by hydroxyapatite can be represented as follows:

$$Ca_{10}(PO_4)_6(OH)_2 + n\,F^- = Ca_{10}(PO_4)_6(OH)_{2-n}F_n + n\,OH^- \qquad (7.18)$$

During the 1940s–1960s, bone charcoal was one of the oldest water defluoridation agents in the United States because of its wide commercial availability; it was also successfully used in many full-scale installations. The exchange capacity of the U.S. Public Health Service (USPHS) plant in Britton, South Dakota, was 102 g fluoride/m^3 bone char in treating waters with initial fluoride concentrations of 5 mg/L.[1,31] The simplicity, local availability, and easy processing facilities made this method more popular for domestic- and community-level applications in many developing countries such as Tanzania. The combined use of bone char with the Nalgonda technique was also reported in literature.[5] However, a major limitation is its poor regeneration capacity, as on many occasions the used bone char gets discarded rather than regenerated. Although the bone char method was successful in removing fluoride to very low levels, it was found to be more expensive and less stable in a continuous flow system than activated alumina (AA). The quality of bone charcoal defines its practical applicability. The bone charring process is very important in this direction, as any failure in this process may result in poor-quality bone charcoal. This may produce drinking waters with bad taste that may smell similar to rotten meat and, ultimately, turn out to be unacceptable to society. Once consumers get exposed to such smell or taste, they may reject the process forever. Also, practical applications of bone char demonstrate that its fluoridation capacity (represented in terms of the quantity of fluoride removed by one gram of bone charcoal at the saturation level) is less than that was being claimed (6 mg/g) in laboratory studies. In actual applications dealing with water treatment, its defluoridation capacity may range from one-third to two-third of this claimed capacity.[16] Of late, bone charcoal defluoridation waterworks are found to be replaced by ion-exchange resins and AA. The domestic-level applications of bone charcoal defluoridation were reported from Tanzania, Thailand, and Africa. In countries such as India, its use was constrained by religious beliefs of many communities. The cost of bone charcoal may vary depending on the method of manufacture. The cost of bone charcoal from the United Kingdom, China, and the United Republic of Tanzania in 1995 was reported to be US$ 2280, US$ 333, and US$ 167 per ton, respectively. However, in many fluorotic areas, it was prepared at a much lower cost; for instance, in Arusha region of Tanzania, by using about 120 kg of charcoal per ton of bone.[16]

7.4.2 Clays and Soils

The potential of clay and soil-based adsorbents for defluoridation has been under investigation in many fluoride endemic regions of the world. Major studies include Illinois soils in the United States, Ando soils in Kenya, sodic soils in India, fired clay chips in Sri Lanka and Africa, and fly ash, Alberta soil, clay pottery, activated clay, kaolinite and bentonite, and illite–goethite soils in China. The Ando soils of Kenya belong to porous soils derived from volcanic ash, which, in part, has weathered to yield "active aluminum" in various forms. This type of soil was reported to have a high adsorption capacity to the tune of 5.51 mg/g.[32] Investigations on improving the adsorption capacity through surface coating of clays and soils were also reported.[33,34] The coating of clays and soils with alumina and iron hydroxides were found to improve their adsorption capacity. Clay, a significant form of soil, has been traditionally used in many developing countries such as India, for making potteries to store water. This clay has been modified with the additions of Al_2O_3, $FeCl_3$ and $CaCO_3$, leading to an observed reduction in water fluoride content.[33] The defluoridation studies on kaolin clay reveal that "solution pH, clay surface area, structure, aluminum content, and the presence of certain exchangeable cations capable of forming fluoride precipitates" are significant in defluoridation.[1,35] Though ion-exchange reactions are believed to be the predominant form of fluoride sorption, they may also be immobilized through the formation of complexes or precipitates with exchangeable cations such as magnesium, iron, and calcium.[1,35] There may be an electrostatic attraction to the clay surface through which fluoride (as F^-) may be retained in the electric double layer.[1,35] It was clearly demonstrated that "disruption of the kaolin crystal structure occurred due to fluoride uptake and F/OH exchange occurred primarily with $Al(OH)_3$ rather than with –OH from the crystal lattice of clay minerals." The fluoride removal mechanism can be represented as follows:[1,35]

$$n(\text{kaolin-OH})_{(s)} + nF^{n-}{}_{(aq)} \leftrightarrow n(\text{kaolin-F})_{(s)} + nOH_{(aq)}{}^{n-} \quad (7.19)$$

The initially sorbed fluoride pushes the layers of metal oxides or hydrated layers on the clay for providing easier access to sorption sites. The hydrogen bonding between kaolin sheets may be disrupted by fluoride by either "attaching themselves to the slightly positive gibbsite surfaces or replacing the hydroxyl groups on these surfaces."[1,36] Due to electronegativity, fluoride ions acquire a highly negatively charge, which will force the silica and gibbsite sheets to further move away from each other. This increases the accessibility and exposure of hydroxyl groups in the gibbsite for fluoride removal. This may enhance an exponential increase in fluoride sorption. The kaolin sheets will be separated to their maximum so that any further separation would not enhance fluoride access to sorption sites. A two-step ligand exchange model was suggested for fluoride sorption onto goethite.[37] Sorption is found to be maximum at around pH 3 in goethite, as fluoride ions hydrolyze and form the neutral species HF near this pH, thus becoming unavailable for sorption.[1,38]

The fluoride-scavenging potential of calcite, quartz, and fluorspar was also investigated. Fluoride uptake was suggested to be a surface adsorption process. The mechanism of fluoride by calcite was considered to work in two phases. In the first phase, calcium ions may get gradually released into the solution at a certain pH range. Further, such dissolved calcium ions interrelate with fluoride ions and form calcium fluoride (CaF_2) precipitates.[8,39] The fluoride replaced CO_3^{2-} from calcite as follows:

$$2\,F^- + CaCO_{3(s)} = CaF_{2(s)} + CO_3^{2-} \tag{7.20}$$

In the initial phase, quartz displayed poor adsorption capacity. The activation of quartz by the ions of iron (Fe^{3+}) drastically increases the fluoride adsorption capacity. It was observed that the siloxane groups of quartz (SiO_2) interrelate with water, forming –SiOH group formulations. The adsorption of fluoride onto quartz is believed to be due to the replacement of F^- for OH^- groups on quartz surfaces.[39] Adsorption capacities of different soils and clays are compared in Table 7.1, which indicates their fluoride adsorption trends.[40] Studies reveal that "hydrated aluminum oxide and iron oxide surfaces occurring in bauxites and goethites/hematites are useful substrates for fluoride sorption."[1,41] Multiple removal mechanisms such as ligand exchange (with surface hydroxyl groups and water molecules), anion exchange, electrostatic attraction, and precipitation are believed to occur.[41] Although applications of different clays for defluoridation are reported in some African countries and Sri Lanka,[40,42] its use in columns is found to be troublesome due to difficulties in packing the columns, controlling the flow, and regenerating the bed. Moreover, in most of the cases, it would not be cost effective.[16] In general, it could be inferred that the clay process would be of either no or, at least, much less use in defluoridation, especially when higher removal efficiencies are expected or higher concentrations of fluoride exist in water.

7.4.3 Carbonaceous and Other Adsorbents

Though attempts were made to attain defluoridation of drinking water by activated carbon, its defluoridation potential was found to be poor. This tendency can be ascribed to the fact that metallic solids such as AA or activated bauxite have an intense affinity for fluoride than nonmetallic solids such as activated carbons. However, it was reported that the adsorption capacity of activated carbon was found to have doubled due to aluminum impregnation, with a maximum capacity of 1.07 mg/g.[43] The carbonized form of the biomass of an aquatic weed Eichhornia crassipes after thermal activation at 600°C demonstrated a removal of 4.4 mg/g.[44] Further, isotherm studies on the algal biomass of *Spirogyra* suggested a maximum adsorption capacity of 1.272 mg/g.[45] The reported adsorption potential of different types of carbon is presented in Table 7.2. It was observed that the fluoride removal mechanism by activated carbon was governed primarily by physical adsorption depending on the specific surface area. Further, the numbers of phenolic hydroxide groups or carboxyl groups on the carbonaceous material surface have no role in fluoride uptake.[30]

TABLE 7.1

Comparison of Fluoride Adsorption Capacity of Major Clay Types

Sorbent Type	Place Description	pH	Initial Fluoride Concentration (mg/L)	Maximum Adsorption Capacity (mg/g)
Gibbsite				
	Australia	5–7	10.0	0.40
	South Africa	5–7	10.0	0.25–0.40
Goethite				
Goethite/Kaolinite	South Africa	5–7	10.0	0.20
Goethite/Illite	China	5–7	10.0	0.23
Goethite/Kaolinite	Sri Lanka	5–7	10.0	0.35
Palygorskite				
Palygorskite/ Dolomite	South Africa	5–7	10.0	0.21–0.29
Smectite	South Africa	5–7	10.0	0.10
	United States	5–7	10.0	Trace
	Alkaline soil, United States	5–7	10.0	0.04–0.08
Kaolinite				
Kaolinite	South Africa	5–7	10.0	0.03
	Acidic soils, United States	5–7	10.0	0.17–0.25
	Acid soils, Illinois	5–7	10.0	0.130
	Pottery clay	5–7	10.0	0.12
	Clay pots, Ethiopia	5–7	10.0	0.07
	South Carolina, Australia	6–7	16–660	4.05
	Kaolin clay	6	10–250	3.48

Source: Coetzee, P.P., Coetzee, L.L., Puka, R. and Mubenga, S., *Water SA*, 29, 331–338, 2003.

The use of hydrous ferric oxide (HFO) for defluoridation revealed that the sorption of fluoride was pH dependent and was taking place by van der Waal's interaction and ion exchange.[46] At an alkaline pH higher than 6, "HFO functions as a cation-exchanger and adsorbs sodium ions present in solution releasing protons,"[1,46] which may reduce the final pH. The maximum adsorption capacity of HFO was found to be 16.50 mg/g. The operating mechanism for fluoride adsorption could be depicted as follows:[46]

TABLE 7.2

Removal of Fluoride by Various Carbonaceous Materials (Initial Fluoride Concentration = 20.0 mg/L, Dose of Adsorbent = 10.0 g/L)

Adsorbent Samples	Base Material	Removal of Fluoride (%)
AC1	Activated carbon (coal)	17
AC2	Activated carbon (wood)	13
AC3	Activated carbon (coal)	12
AC4	Activated carbon (wood)	16
AC5	Activated carbon (wood)	6.6
AC6	Activated carbon (petroleum coke)	5.4
CB	Carbon black	10
CC 1	Charcoal	3.5
CC 2	Charcoal	3.2
CC 3	Charcoal	1.6
CC 4	Charcoal	0.40

Source: Abe, I., Iwasaki, S., Tokimoto, T., Kawasaki, N., Nakamura, T. and Tanada, S., *J. Colloid Interface Sci.*, 275, 35–39, 2004.

In the pH range of 2.0–5.0:

$$Fe_2O_3 \cdot xH_2O_{(solid)} + F_{(aq)}^- \leftrightarrow Fe_2O_3 \cdot (x-1)H_2O \cdot H^+F_{(solid)}^- + OH_{(aq)}^- \quad (7.21)$$

and for pH > 6:

$$Fe_2O_3 \cdot xH_2O_{(solid)} + Na^+_{(aq)} + F^-_{(aq)} \leftrightarrow Fe_2O_3 \cdot (x-1)\ H_2O \cdot OH^-Na^+F^-_{(solid)} + H^+_{(aq)} \quad (7.22)$$

Adsorbents such as Chitin (refers to the polysaccharides commercially extracted from shellfish processing wastes and naturally available in organisms such as bacteria, fungi, etc.), its deacetylated product chitosan (mainly obtained from crustacean shells of prawn, crab, shrimp, or lobster), and 20% lanthanum-incorporated chitosan were tested in removing excess fluoride from drinking water.[47] It was observed that the 20% La-chitosan showed a much better fluoride removal capacity with a higher uptake of 3.1 mg/g at an adsorbent dose of 1.5 g/L.[47] However, the capacity reduced drastically in treating natural water (nearly reduced to half) compared with synthetic systems. In experiments dealing with magnetic-chitosan particles, adsorption capacities of 3–17 mg/g were reported as corresponding to initial fluoride concentrations of 5–140 mg/L.[48] The Langmuir-saturated monolayer capacity of Laterite, the geomaterial, was found to be 0.8461 mg/g with an experimental maximum column capacity of 0.3473 mg/g.[49] Of late, MgAl-CO_3-layered double hydroxides (LDHs) were employed to treat high fluoride concentration solutions.[50] LDHs (known as hydrotalcite-like materials) are a class of naturally occurring and synthetic anionic clays, and they display a maximum adsorption capacity of 319.8 mg/g in isotherm studies.[50]

7.4.4 Alumina

The surface structure and adsorption play an important role in the use of alumina as a catalyst and its application in separation processes. The acidity–basicity properties of alumina are found to be the major determinants in defining its adsorption behavior. Water may be either physisorbed or chemisorbed onto alumina surfaces. However, the amount of sorption depends on temperature and vapor pressure. Water may be adsorbed as undissociated molecules with strong hydrogen bonds at ambient temperatures. Further, at slightly elevated temperatures, surface hydroxyl groups are formed that will be expelled as H_2O at further higher temperatures. "Chemisorption of water onto alumina surfaces is generally regarded as a Lewis acid–base adduct formation with the Al^{3+} ion acting as an electron pair accepter (Lewis acid) and the hydroxyl ion acting as an electron pair donor (Lewis base). The hydroxyl groups on the alumina surface are sources of protons and therefore behave as Bronsted acid sites. But oxygen bridges formed on the alumina surface through dehydration of two adjacent OH^- groups are active Lewis acid sites."[1,51,52] Knozinger's model is regarded as a total advancement for understanding the OH surface groups on alumina. It is assumed that the "termination of alumina crystallites occurs along three possible crystal planes. Depending on the coordination properties of surface anions and the number of Al ions attached to hydroxyl groups," five types of hydroxyl groups may be present on the three possible crystal planes.[1,53]

The acidity of the hydroxyl groups as well as the OH-stretching frequency is affected by the net charge of alumina. "Hydroxyl groups with highest frequency possess the highest basicity and those with the lowest frequency are thought to possess the highest acidity. The surface hydroxyl groups of hydrated alumina are amphoteric in nature and therefore may ionize as Bronsted acids or bases depending on the circumstances. This ionization is responsible for the surface charging at the aqueous interface of alumina"[1] and is believed to be generated through a two-step process. The first step in this direction involves "surface hydration." This is an effort by the exposed surface atoms for completing their respective coordination shells. Aluminum cations achieve surface hydration by linking to an OH^- ion or a water molecule. However, oxygen ions pull out a proton from water. The second step is the dissociation of the surface hydroxyls. Both the steps create surface hydroxyls that can ionize as Bronsted acids or bases, giving rise to surface charges.[1]

It is plausible that alumina surfaces may get charged by specific adsorption of ions other than protons. However, it is suggested that pH has considerable influence on the nature and amount of the surface charge of alumina. "The point of zero charge (pH_{ZPC}) is the pH value at which net surface charge is zero. In acidic medium below pH_{ZPC}, the hydroxyl groups on the surface are protonated and therefore the surface has a net positive charge. At basic pH, the surface is negatively charged as the hydroxyl groups act as Bronsted acids and give off protons."[1,51] This "surface charging" in alumina is instrumental in the formation of "an electrical double layer" at the aqueous interfaces due to electrostatic interactions between the charged surfaces and oppositely charged ions in the bulk solution. It could be inferred that this pH-dependent "surface charge" and associated "electrical double layer" formation at aqueous interfaces turn out to be instrumental in removing impurities from water.

The adsorption on alumina from aqueous solutions can be expressed in terms of two models, namely, the ligand-exchange model and the ion-exchange model. The ability of surface hydroxyls to dissociate or to be protonated (depending on the pH) defines the ion-exchange properties, as shown by Equations 7.23 and 7.24[52] as follows:

$$\text{M-OH} + H^+ \leftrightarrow \text{M-OH}_2^+ \tag{7.23}$$

$$\text{M-OH} + OH^- \leftrightarrow \text{M-O}^- + H_2O \tag{7.24}$$

The ability of alumina to exchange ligand originates from the "presence of Lewis acid sites on the surface (Al^{3+}) and the presence of water molecules bonded to the sites which can be exchanged for other Lewis based molecules." The following equilibria have been used to describe the ligand-exchange reactions by Equations 7.25 through 7.29) as follows:[52,54,55]

$$M(OH)(H_2O) + L_1^- \leftrightarrow M(H_2O)L_1 + OH^- \tag{7.25}$$

$$M(OH)(H_2O) + L_1^- \leftrightarrow M(OH)L_1^- + H_2O \tag{7.26}$$

$$M(OH)(H_2O) + L_2^- \leftrightarrow M(H_2O)L_2 + OH^- \tag{7.27}$$

$$M(OH)(H_2O) + L_2^- \leftrightarrow M(OH)L_2^- + H_2O \tag{7.28}$$

$$M(OH)L_1^- + L_2^- \leftrightarrow M(OH)L_2^- + L_1^- \tag{7.29}$$

M represents the metal oxide metal, and L_1 and L_2 represent Lewis bases present in water and solute, respectively. Equations 7.23 and 7.24 describe the modifications of the surface sites by Lewis bases present in water, whereas Equations 7.25 through 7.29 show ligand exchange with a harder base from the solute (L2). The "driving force for Ligand exchange in alumina is the affinity Al^{3+} (a hard Lewis acid) has for hard Lewis bases."[1]

7.4.5 Activated Alumina

AA is a granular porous material with a very high surface area (200–300 m^2/g); it consists mainly of aluminum oxide (Al_2O_3). Of late, the application of AA in the defluoridation of drinking water is widely accepted worldwide.[56–59] The sorption mechanism of AA could be effectively described by the surface complex formation model (ligand-exchange model). This model signifies the active role of surface hydroxo groups in surface complex formation. It is suggested that amphoteric hydroxo groups develop on the alumina surface due to hydration, as shown by Equations 7.30 and 7.31 as follows:[21,25]

$$AlOH_2^+ = AlOH + H^+ \tag{7.30}$$

$$AlOH = AlO^- + H^+ \tag{7.31}$$

where AlO^-, $AlOH_2^+$ and AlOH are, respectively, the negative, positive, and neutral surface hydroxo and oxo groups. The adsorption model of fluoride can be described by Equations 7.32 through 7.34[21] as follows:

$$AlOH + F^- = AlF + OH^- \quad (pH > 7) \tag{7.32}$$

As represented in the earlier equation, OH^- is released from AA surface into the bulk phase, resulting in an increase in pH. The extent of OH^- release is the most pronounced at $pH > 6–7$ and diminishes at $pH < 7$.[25,60,61] However, at a lower range of pH (3–5), a different adsorption mechanism may be in operation as follows:

$$AlOH_2^+ + F^- = AlF + H_2O \quad (pH < 6) \tag{7.33}$$

At low pHs, it is plausible that the amount of fluoride adsorbed may exceed the total available surface sites during higher initial fluoride concentrations. This could be represented by a polynuclear surface complex formation as follows:[21]

$$AlOH + 2F^- = AlF_2^- + OH^- \tag{7.34}$$

In the presence of fluoride, AA is rendered soluble by forming alumina-fluoro complexes that are stable in acidic pH and may turn unstable with a rise in pH. So, AA sorption systems for defluoridation should be carried out at pH values where alumina-fluoro complexes are unable to resist alumina from dissolving. The regeneration of AA is usually carried out with a caustic solution (usually 1% sodium hydroxide) followed by a dilute acid (usually 0.05 N sulfuric acid) and water rinse.[1,62] The following reactions take place during regeneration:

$$AlF\,(s) + OH^- \rightarrow AlOH\,(s) + F^- \tag{7.35}$$

$$AlOH\,(s) + H_2SO_4 \rightarrow AlHSO_4\,(s) + H_2O \tag{7.36}$$

The characteristic features of AA, excellent scavenging potential, and specific affinity for fluoride make it an ideal candidate for defluoridation. Packed beds of granular AA have been traditionally used for defluoridation of public water supplies, and their reported capacity is in the range of 6,750–11,760 g/m^3 in a continuous flow system. The minimum interference from counter-ions and the attractive costs are added advantages for AA-based systems.[1] The Department of Science and Technology, Government of India, has supported the Public Health Engineering Departments of State Governments of fluoride endemic areas to develop hand pump–attached defluoridation units. A cylindrical defluoridation unit fabricated in this direction and field tested in Makkur, Unnao district of the state of Uttar Pradesh, in 1993 had been reported by Daw (2004).[56] The AA (particle size 0.3–0.9 mm) bed of 110-kg alumina in 55-cm height is packed in a mild steel cylindrical drum that is 0.5 m in diameter and 1.5 m in height. An elevated hand pump discharge level and elevated platforms are provided. A bypass line was provided to draw water directly from the hand pump for uses other than drinking and cooking. A system was designed for a raw water fluoride concentration of 6–7 mg/L. The two important parameters that define the field application of AA for defluoridation are its fluoride uptake capacity and reuse potential.[1,56,58]

The application and adaptation of AA as a defluoridating medium with a hand pump or DDUs started gaining ground in India. The United Nations Children's Fund (UNICEF) and Rajiv Gandhi Drinking Water Mission (RGDWM) are taking leadership in this front. The DDU consists of two chambers. The upper chamber, fitted with a micro-filter and an orifice to give a flow rate of about 12 L/h, is charged with 3–5 kg of AA packed to a depth of 9–17 cm. The perforated stainless-steel plate on the top of the AA bed ensures uniform distribution of raw water. A lid is provided at the upper chamber, and a tap is provided at the bottom. The raw water filled in the upper chamber percolates through the adsorbent bed, and the fluoride in the influent gets adsorbed by the AA media. Treated water can be collected in the lower chamber, and it can be drawn as and when required.[28,56]

Further, novel and user-friendly techniques for the regeneration of exhausted AA were developed. In UNICEF-assisted projects, the spent AA beds and sludge are used for making bricks. The practical adaptability and technical viability of AA in small community-level applications, such as household DDUs and water taps, has been successfully demonstrated[56,58,63] in many countries such as India, China, and Thailand. However, international performance standards are yet to be framed.[16] Further, AA poses some limitations. Regeneration brings in a reduction of about 5%–10% in material, 30%–40% in capacity and an increased presence of aluminum (>0.2 mg/L) in the effluent. It was suggested that normal pHs may keep the aluminum residual within permissible limits. Though adjustment of pH can be done at waterworks, actual pH of the raw water is to be relied on in domestic and small community treatments. So, the fluoride-scavenging potential of AA has to be verified through field testing under genuine ground-level conditions. Though a high fluoride scavenging potential of AA (to the range of 4–15 mg/g) has been reported in literature,[21] field experiences demonstrate a sharp reduction in its scavenging potential (of around 1 mg/g).[16]

7.4.6 Other Alumina-Based Adsorbents

Metal oxyhydroxides possess surface oxygen that is different from the number of coordinating metal ions. This felicitates the adsorption of different anions and cations by oxyhydroxides. Successful application of activated bauxite for defluoridation turns significant in this direction.[25] Red mud, the bauxite waste of alumina manufacturing, is an unwanted by-product of alkaline leaching of bauxite that is also used for defluoridation.[64] The adsorption of fluoride is found to be maximum at pH 5.5. The observed sharp reduction in fluoride removal at a pH above 5.5 could be ascribed to the stronger competition of fluoride with hydroxide ions on the adsorbent surface. The possibilities of the formation of weakly ionized hydrofluoric acid in a high acidic range may also reduce adsorption. The use of refractory-grade bauxite for fluoride removal[65] demonstrates the significance of pH. Alum sludge (a waste product obtained during the manufacture of alum from bauxite through sulfuric acid process) was found to have a fluoride adsorption capacity of 5.394 mg/g.[24] It was also found that thermal activation at moderate temperatures (300°C–450°C) increases the adsorption capacity of titanium-rich bauxite.[66]

Amorphous alumina-supported carbon nanotubes (CNTs) were reported successful for defluoridation. The adsorbent was prepared by heating the composites of $Al(NO_3)_3$ and CNTs at 500°C for 2 h under N_2 atmosphere. It was observed that the best fluoride adsorption on Al_2O_3/CNTs occurs at a wide pH range of 5.0–9.0. The isoelectric point (IEP) of Al_2O_3/CNTs is found to be 7.5. At $pH < 7.5$, the surface charge of the material is positive, and, consequently, columbic attraction may trigger an interaction between the adsorbent and fluoride ions. At $pH \geq IEP$, due to the neutral or negatively charged nature of the adsorbent, its scavenging potential may get reduced. The excellent fluoride adsorption capacity exhibited by Al_2O_3/CNTs was also attributed to the amorphous structure of Al_2O_3. The removal of fluoride by CNTs may be due to their large surface area or surface reactions, resulting in the formation of functional groups by oxidation. At pH 6, Al_2O_3/CNTs exhibit a fluoride adsorption capacity of 28.7 mg/g at equilibrium for a fluoride concentration of 50 mg/L, which is 13.5 times higher than that of alumina-impregnated carbons.[1,67]

Rare earth oxide-based mixtures and rare earth element impregnation of porous adsorbents or carrier materials have shown promising results in defluoridation. In this direction, the use of lanthanum-impregnated silica gel is significant, as it is found to have an adsorption potential of 3.8 mg/g. The chemical adsorption reaction between lanthanum and silica gel can be explained as follows:[9]

$$-Si-OH \rightarrow -Si-O^{-} + H^{+} \tag{7.37}$$

$$-Si-O^{-} + La(OH)_2^{+} \rightarrow -Si-O-La(OH)_2 \tag{7.38}$$

It was observed that fluoride removal by the original silica gel was negligible in a wide pH range (4–10). This signifies the active role played by adsorbed lanthanum in defluoridation. In general, the mechanism for the removal of anions by chemical adsorption by lanthanum-impregnated silica gel can be explained as follows:[9]

$$-Si-O-La(OH)_2 + A^{(z-n)} \rightarrow -Si-O-La(OH)\,A^{(z-n)} + n\,OH^{-} \tag{7.39}$$

where $A^{(z-n)}$ is the anion, for instance, fluoride, phosphate, and arsenate ions. The observed pH rise (from 3.5 to 7.1) is due to the release of OH^- ions from the adsorbent, as shown in Equation 7.39. However, an adsorbent developed by impregnating lanthanum on cross-linked gelatin has been reported to have a high fluoride adsorption potential of 21.28 mg/g.[68] Gelatin is a polypeptide with many functional groups. These functional groups have strong attraction for metal ions. Glutaraldehyde (GTA) is used for cross-linking of gelatin, which involves the "reaction of free amino groups of lysine or hydroxylysine amino acid residues of the polypeptide chains with the aldehyde groups of GTA to form a Schiff's base."[1,69] Impregnation was carried out by using $La(NO_3)_3$ solution at different pH. It was observed that a pH range of 5–8 is optimum for La^{3+} impregnation of cross-linked gelatin, indicating the principal role of carboxyl groups of protein in binding. The removal mechanism of fluoride and the regeneration process of lanthanum-impregnated cross-linked gelatin can be represented[1,70,71] as shown in Figure 7.2.

The fluoride-scavenging characteristics of zeolite F-9 of size ranging from 0.15 to 0.30 mm containing surface-active sites created by exchanging Na^+-bound zeolite with Al^{3+} or La^{3+} ions was investigated. It was observed that the three-dimensional skeletal structure of zeolite has small pores in which the exchangeable cations are located, facilitating the mechanisms of ion exchange. The exchange of the trivalent ions La^{3+} and Al^{3+} for Na^+ ions attached to the zeolite F-9 can be represented as follows:[72]

$$Me^{3+}(soln) + 3\,Na^{+}(zeo) \leftrightarrow Me^{3+}(zeo) + 3\,Na^{+}(soln) \tag{7.40}$$

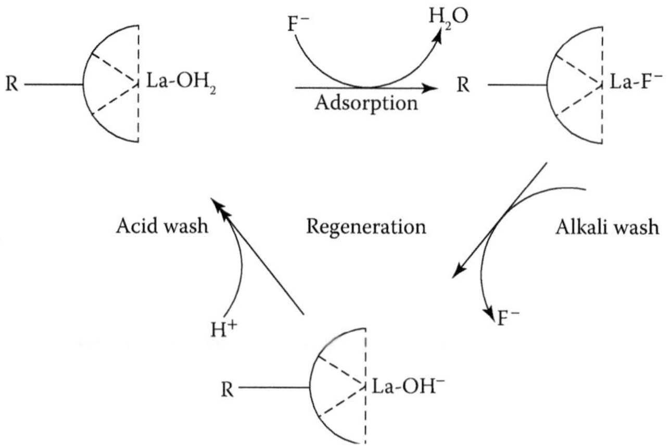

FIGURE 7.2
The adsorption and regeneration mechanism of lanthanum-impregnated cross-linked gelatin. (From Zhou, Y., Yu, C. and Shan, Y., *Sep. Pur. Technol.*, 36, 89–94, 2004.)

where Me^{3+} refers to La^{3+} or Al^{3+} and (soln) and (zeo) denote the solution and zeolite phases. The porosimetric studies (done on samples after sorption) revealed an increase in porosity of zeolite particles from 25% to 32%, where Al^{3+} was loaded onto zeolite. When La^{3+} was loaded onto zeolite, the corresponding increase was 38%. It was also pointed out that at around neutral pH, the surface of zeolite turns heterogeneous as the trivalent metals used for surface modification of zeolite form many complexes (protonated and nonprotonated). At equilibrium, the pH was found to increase from 4.0 to 6.86 in Al^{3+}-exchanged zeolites. For La^{3+}-exchanged zeolites, the increase was from 4.0 to 6.36, as depicted by the following equations:[72]

$$\text{Zeo-MeOH}_2^+ + \text{OH}^- \leftrightarrow \text{Zeo-MeOH} + \text{H}_2\text{O} \tag{7.41}$$

$$\text{Zeo-MeO}^- + \text{H}^+ \leftrightarrow \text{Zeo-MeOH} \tag{7.42}$$

For fluoride sorption on aluminol surface sites (Me = Al):

$$\text{Zeo-MeOH}_2^+ + \text{F}^- \leftrightarrow \text{Zeo-MeF} + \text{H}_2\text{O} \tag{7.43}$$

$$\text{Zeo-MeOH} + \text{F}^- \leftrightarrow \text{Zeo-MeF} + \text{OH}^- \tag{7.44}$$

The interaction between La-exchanged zeolite and fluoride can be represented as (Me = La) follows:

$$\text{Zeo-MeOH}_2^+ + \text{F}^- \leftrightarrow \text{Zeo-MeOH}_2^+ \ldots \text{F}^- \tag{7.45}$$

where Zeo-MeOH, Zeo-MeO^- and Zeo-$MeOH_2^+$ are the neutral, hydroxylized, and protonated surface sites of zeolite, respectively. The expression "Zeo-Me" denotes the zeolite–metal surface, where Me represents either La or Al. As shown, the sorption of fluoride on aluminol sites is mainly by ion exchange and inner-sphere complexation, as expressed by Equations 7.41 through 7.44. As shown in Equation 7.45, the mechanism of fluoride sorption in La-exchanged zeolite is of a physical nature, demonstrating the formation of an outer-sphere complex. Adsorption capacities of 40–43 mg/g and of 42.0–58.5 mg/g were reported for Al^{3+}-exchanged zeolite and La^{3+}-exchanged zeolite, respectively.[72] Alum-impregnated activated alumina (AIAA) was also experimented for defluoridation.[73] AIAA displayed a highly pH-dependent removal mechanism with adsorption capacities of 192.65, 40.68 and 19.80 mg/g at pH values of 4.0, 6.5 and 9.0, respectively.[73]

The use of ion-exchange resins are also found to be effective in removing fluoride from water. However, during the removal by ion-exchange resins, it is most likely that similar anions in water will also be removed. So, fluoride exchange capacity mainly depends on the ratio of fluoride to the total anions in solution. It is naturally expected that the exchange capacity of resins increases swiftly with higher values of fluoride to total anion ratios. However, it is reported that most fluoride waters have comparatively low fluoride concentrations. In addition to the relatively lower value of exchange capacity, the high cost of resins compared with other adsorbents such as AA should also be considered. This limits the use of synthetic anionic resins. Further, since the removal also includes the sorption of other anions, the sorption capacity of resins appears too low to the tune of 0.5 mg/L.[74] A study on defluoridation by a two-way ion-exchange cyclic process using two anion-exchange columns was also reported.[11] The results show that the two-way ion-exchange cyclic process may remove fluoride within the range limits of the defluoridation potential of anion-exchange resins. A detailed analysis of the adsorbents so far used in defluoridation clearly suggests that a universally adaptable defluoridation medium is still an elusive goal. Given the magnitude and diversity of the problem, the choice mainly depends on the availability of the adsorbent adjacent to a fluorotic endemic area, that too at reasonable, affordable, or no cost.

7.5 Electrochemical Methods

7.5.1 Electrocoagulation

Electrocoagulation (EC) employs an electrolytic process for producing a coagulant in situ through the oxidation of appropriate anodic materials. The coagulant ions thus released react with the ions (targeted to be removed), initiating a normal coagulation process. Generally, aluminum and iron are used as sacrificial electrodes that produce ions continuously in the system. Additionally, these sacrificial electrodes protect the normal anode by reducing its dissolution potential and the cathode through passivation. When an electric current passes through the Al electrodes, an anodic reaction releases Al ions that react with OH ions produced at the cathode and with F ions in solution. As a result of electro-condensation, the defluoridation efficiency of the EC system may be higher than the traditional coagulation process. Since F^- ions are attracted to the anode, the concentration of F^- ions near the anode exceeds that in the bulk solution, leading to higher efficiency through the "condensation effect" (Figure 7.3).[10]

Of late, the EC process was modified through the introduction of bipolar electrodes.[75] For developing bipolarity, a conductive plate without any electric connection is placed between two electrodes having opposite charges, as shown in Figure 7.4. The anodic reaction will commence at the positive side of the bipolar electrode, and cathodic reactions will commence at the negative side. The Al^{3+} ions (produced by the dissolution of Al anode) at appropriate pHs get initially transformed to aluminum hydroxide $Al(OH)_3$ and ultimately polymerized to $Al_n(OH)_{3n}$, which have a very strong affinity toward F^- ions as follows:[76]

$$Al \rightarrow Al^{3+} + 3e^- \quad (7.46)$$

$$Al^{3+} + 3\,H_2O \rightarrow Al(OH)_3 + 3H^+ \quad (7.47)$$

$$nAl(OH)_3 \rightarrow Al_n(OH)_{3n} \quad (7.48)$$

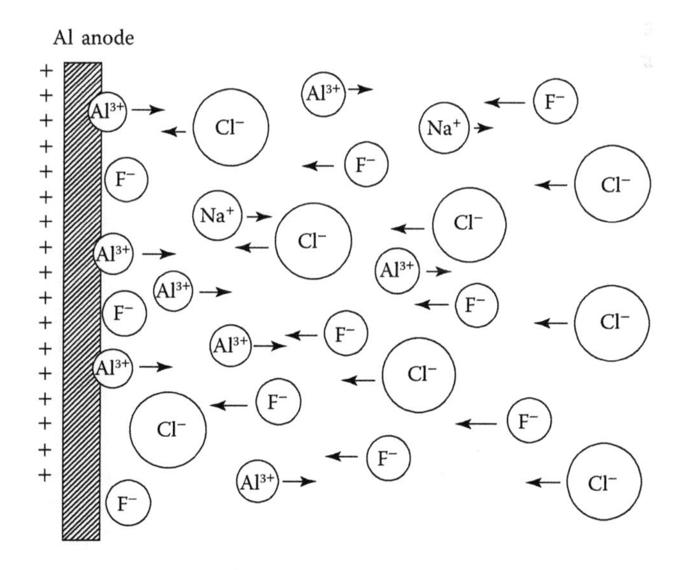

FIGURE 7.3
Diagram representing the electro-condensation effect. (From Hu, C.Y., Lo, S.L., and Kuan, W.H., *Water Res.*, 37, 4513–4523, 2003.)

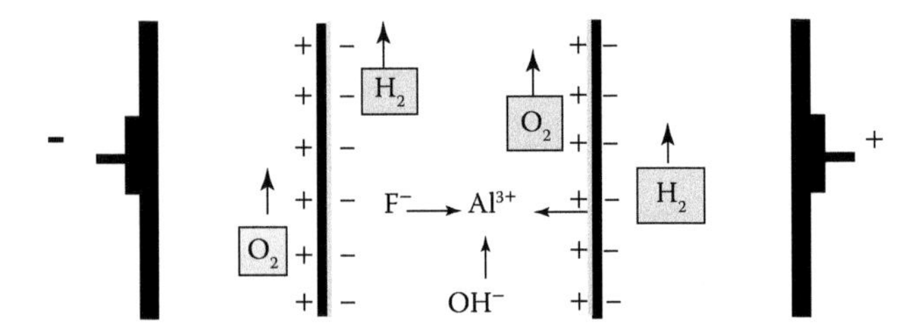

FIGURE 7.4
Schematic diagram of bipolar electrode system. (Modified from Mameri, N., Yeddou, A.R., Lounici, H., Belhocine, D., Grib, H. and Bariou, B., *Water Res.*, 32, 1604–1612, 1998.)

It could also be plausible that the Al^{3+} ions under high fluoride concentrations at the anode may be induced to form AlF_6^{3-} ions (Equation 7.49). These may get transformed into an insoluble salt (Na_3AlF_6) by sodium ions (Equation 7.50).[76]

$$Al^{3+} + 6F^{-} \rightarrow AlF_6^{3-} \quad (7.49)$$

$$AlF_6^{3-} + 3Na^{+} \rightarrow Na_3AlF_6 \quad (7.50)$$

The reactions encountered at the cathodic (Equation 7.51) and anodic compartments (Equation 7.52) are as follows:

$$2\,H_2O + 2e^{-} \rightarrow H_2 + 2OH^{-} \quad (7.51)$$

$$2OH^{-} \rightarrow \tfrac{1}{2}O_2 + H_2O + 2e^{-} \quad (7.52)$$

The efficiency of this process was nearly 100% at optimum pH (5–7.6). However, a rise in temperature may reduce the efficiency due to fluoride desorption from aluminum hydroxide and/or by destruction of fluoro-aluminum complexes. Recently, the concept of hydro-fluoro-aluminum complex $Al_n(OH)_m F_k^{3n-m-k}$ has been introduced to describe various chemical species containing fluoride, hydroxide, and Al^{3+}, such as Al-F complexes, Al-OH complexes and the hydro-fluoro-aluminum colloid flocs.[77] The electrical field encourages $Al_n(OH)_m F_k^{3n-m-k}$ to be condensed near the electrodes, making fluoride sorption mostly on the electrodes and resulting in a higher defluoridation efficiency in the EC process.[1] However, issues such as interference from other anions present in water due to the competition effect may reduce the efficiency of the EC process. For example, the presence of sulfate may significantly reduce the efficiency of the process due its strong affinity toward Al^{3+} ions. Further, regular replacement of sacrificial electrodes due to continuous dissolution into solution because of oxidation and high consumption of electric power during operation may also be treated as limitations of the process.

7.5.2 Electrosorption

Electrosorptive techniques could be used to enhance sorption capacity of the conventional systems. It was demonstrated that the efficiency of the alumina bed got significantly enhanced through a new activation technique by the application of an electric field. An electrochemical cell is prepared with two stainless-steel electrodes. These electrodes were introduced into a 20-cm-long and 2-cm-diameter PVC column to produce an electrical field in the AA bed. This technique exhibited more efficiency than the conventional activation techniques. The regeneration of an adsorbent plays a crucial role in its selection, economical evaluation, and field application. Any reduction in the cost of the regeneration will significantly add to the efficiency of the adsorbent bed and the process.[78] It was observed that during regeneration, the electrical field created between two electrodes gave greater mobility to OH ions in accessing active sites located within the pores, thereby improving the regeneration of AA. Successive regeneration in three cycles could be done without any reduction in the sorption capacity of the used adsorbent. Moreover, the volume of the cleaning agent (NaOH) used for regeneration got drastically reduced. Further, more than 95% recovery of the adsorption capacity was ensured with electrosorption techniques. It is also pointed out that the amount of water necessary for the regeneration of the saturated bed was very minimal compared with conventional techniques.[1,78]

7.6 Membrane Processes

Membrane may be treated as a selective barrier controlling material transport between two adjacent phases. Thus, it could be regarded as an interphase between the two phases. So, in separation science, the main advantage of membrane technology is related with the transport selectivity of the membrane.[79] A wide range of technical applications are possible by altering or changing the membrane structure, cross section, and shape. The difference in chemical potential due to a concentration or pressure gradient across the membrane (or by an electric field) acts as the driving force for passive transport in membrane separations. Depending on the membrane barrier structures and the trans membrane gradients, the process is characterized as shown in Table 7.3.

Though different membrane technologies (Table 7.3) have been used over the years in diverse fields, their application in removing inorganic anions such as fluoride is of recent origin. The applications of membrane processes turn attractive in water treatment, as many of the difficulties associated with precipitation, coagulation, or adsorption can be avoided or minimized. Recently, many hybrid systems have been developed in which one membrane

TABLE 7.3

Classification of Membranes and Membrane Processes for Separations via Passive Transport

		Trans Membrane Gradient		
S. No.	**Membrane Barrier Structure**	**Concentration**	**Pressure**	**Electrical Field**
1	Nonporous		Reverse osmosis (RO)	Electrodialysis (ED)
2	Microporous pore diameter, $d_p \leq 2$ nm	Dialysis (D)	Nanofiltration (NF)	
3	Mesoporous pore diameter , d_p = 2–50 nm	Dialysis (D)	Ultrafiltration (UF)	Electrodialysis (ED)
4	Macroporous pore diameter, d_p = 50–500 nm		Microfiltration (MF)	

Source: Ulbricht, M., *Polymer*, 47, 2217–2262, 2006.

process has been integrated with another to produce even higher-quality water.[6] The basic principles governing these membrane-based processes are discussed next.

7.6.1 Reverse Osmosis

Reverse osmosis (RO) is a pressure-driven membrane process. In this process, the driving force for transport across the membrane is the pressure gradient between the water to be treated and the permeate side. The most popular application of RO is in desalination of drinking water and process water. The applied operating trans membrane pressure ranges from 20 to 100 bar.[6] The type of polymers used and the morphological details of the barrier are given in Table 7.4. The application of RO is similar to the ordinary filtration process. The major difference is in the role of osmotic pressure. The osmotic pressure is relatively insignificant in ordinary filtration; however, it is a crucial factor in RO. The RO modules of spiral-wound polyamide membranes were successfully employed in treating electronic industry wastewater. This wastewater has an average fluoride concentration around 400 mg/L and is generated due to the chemical cleaning of silicon plates with a diluted hydrofluoric acid.[80] It was reported that a bench-scale RO unit with a capacity of 0.15 m^3/h removed more than 60% of fluoride from feed water concentrations of 1.7 mg/L employing spiral-wound cellulose acetate membranes.[81] The implementation of a pilot scale RO plant in Lakeland, California, was also reported successful in treating fluoride concentration in the range of 3.6–5.3 mg/L.[82]

The major limitations of RO include the high energy consumption needed for maintaining the required pressure difference. In addition to

TABLE 7.4

Polymer-Based Separation Membranes for Different Processes

		Morphology		
Membrane Process	**Polymer**	**Barrier Type**	**Cross Section**	**Barrier Thickness (μm)**
Reverse osmosis	Cellulose acetates	Nonporous	Anisotropic	~0.1
	Polyamide, aromatic, in situ synthesized	Nonporous	Anisotropic/ composite	~0.05
	Polyether, aliphatic cross-linked, in situ synthesized	Nonporous	Anisotropic/ composite	~0.05
Nanofiltration	Polyamide, aromatic, in situ synthesized	Nonporous	Anisotropic/ composite	~0.05
	Polyether, aliphatic cross-linked, in situ synthesized	Nonporous	Anisotropic/ composite	~0.05
	Polyimides	Nonporous	Anisotropic	~0.1
	Polysiloxanes	Nonporous	Anisotropic/ composite	~0.1<1–10
Ultrafiltration	Cellulose acetates	Mesoporous	Anisotropic	~0.1
	Cellulose, regenerated	Mesoporous	Anisotropic	~0.1
	Polyacrylonitrile	Mesoporous	Anisotropic	~0.1
	Polyetherimides	Mesoporous	Anisotropic	~0.1
	Polyethersulfones	Mesoporous	Anisotropic	~0.1
	Polyamide, aromatic	Mesoporous	Anisotropic	~0.1
	Polysulfones	Mesoporous	Anisotropic	~0.1
	Polyvinylidenefluoride	Mesoporous	Anisotropic	~0.1
Electrodialysis	Perfluorosulfonic acid polymer	Nonporous	Isotropic	50–500
	Poly(styrene-co-divinylbenzene), sulfonated	Nonporous	Isotropic	100–500

Source: Ulbricht, M., *Polymer*, 47, 2217–2262, 2006.

biological fouling of the membrane (due to natural organic matter [NOM] and microorganisms), mineral fouling may also occur due to precipitation of certain salts. Further, interference from divalent metallic cations such as Ca^{2+} and Mg^{2+}, and anions such as SO_4^{2-} may interfere with the separation of particular toxic anionic species. This may affect water recovery and reduce the osmotic pressure[83] demanding pre treatment of raw water. At lower pH values, because of strong hydrogen bonding of fluoride in acidic solution, its permeability and solubility in the membrane may increase. This will drastically reduce the life of the membrane and force its replacement.[84] It was reported that while treating fluoride-rich waters, up to 99%

of the salts in water were getting rejected by the membrane.[85] This may result in unpleasant taste, as treated water may lack the right balance of minerals.[1,86]

7.6.2 Nanofiltration

NF is a membrane-separation process that is targeted to remove uncharged organic species on nanoscale by size exclusion, and ions by charge effects.[87] UF is a membrane processes targeting macromolecular substances. RO, UF, and NF are pressure-driven membrane processes targeting removal of multivalent ions from monovalent species. Through judicious selection of membranes and operating conditions, ions of the same valence can also be separated. NF membranes are essentially low-pressure RO membranes that are capable of removing hardness along with a wide range of organic (bacteria, viruses, and pesticides) and inorganic components (nitrates, arsenic, and fluoride) in a single process.[1,87] NF membranes provide higher water fluxes at lower trans membrane pressures[6] than conventional RO membranes. So, NF membranes are commonly referred to as *low-pressure RO membranes*.[6] The applied pressure gradient across the membrane paves the way for solute transport by convection. Given a negative charge in a neutral and alkaline environment, NF membranes provide an asymmetric pattern (Table 7.4). The surface charge on NF membranes is due to the anions adsorbed onto their surface. In contrast, the surface charge in ion-exchange membranes is due to various fixed charged groups that are bonded to the polymer structure.[6,88] In addition to diffusion and convection mechanisms, the repulsion between anions is also significant. These repulsive actions between surface groups will be higher in the case of multivalent anions.[6,87] In addition to size-based exclusion, the system offers mechanisms of ion exclusion as well. So, relatively higher degrees of ion rejections or separations at higher water fluxes, as in the case of RO, could be achieved through the NF membranes.[1] Of late, one of the world's largest NF plants is operative in Paris, France, with a capacity of 140,000 m^3/d and having more than 9,000 Filmtec NF200 membrane modules.[86] Another NF plant in Finland (capacity of 380–600 m^3/d) was reported successful in fluoride removal to the tune of 76% with NF 255 (Filmtec).[89] Negatively charged commercial thin-film composite (TFC) membranes were also used for fluoride removal.[85] Most recently, hybrid systems using NF in combination with subsequent adsorption or biodegradation were also attempted.[6] Recent trends in membrane development encompass the applications of existing RO thin-film composite membranes to the usage of NF for water treatments. The replacement of multiple treatment processes and techniques by single-membrane applications is the need of the hour. Currently, this triggers advanced research in NF.

This trend is gaining ground as one of the best available technologies in many developed countries. Though NF is a preferred option and an increasingly recommended technique globally as against the traditional RO systems, it is constrained with limitations as well. Among them, the treatment and management of relatively large concentrations of retentate fraction (generated to the tune of nearly 20% of the feed volume) is a matter of concern.

7.6.3 Electrodialysis

Electrodialysis (ED) is a process in which the transport of ions present in wastewater is accelerated due to an externally applied electric potential difference. ED has been conventionally used for desalination and demineralization of brackish waters. In an ED cell, cation- and anion-selective membranes are placed in a parallel fashion across the current path[90] (Figure 7.5). By introducing a current, cations move through the cation-exchange membrane toward the cathode and anions move through the anion-exchange membrane to the anode. As a result, salinity may decrease in one space in the alternating spaces between the membranes. However, salinity may increase in the next space throughout the stack of parallel membranes. The desired level of salinity could be achieved once water is made to pass through several such stacks. The cost of ED is directly proportional to the salinity of the water to be treated. As in RO, ED is also bound with limitations of membrane fouling. This may largely be ascribed to the presence of colloidal elements and the precipitation of sparingly soluble salts such as ferric hydroxides, calcium sulfates, and calcium carbonates. Most negatively charged colloids present in natural waters may get deposited on the anion-exchange membranes.[6]

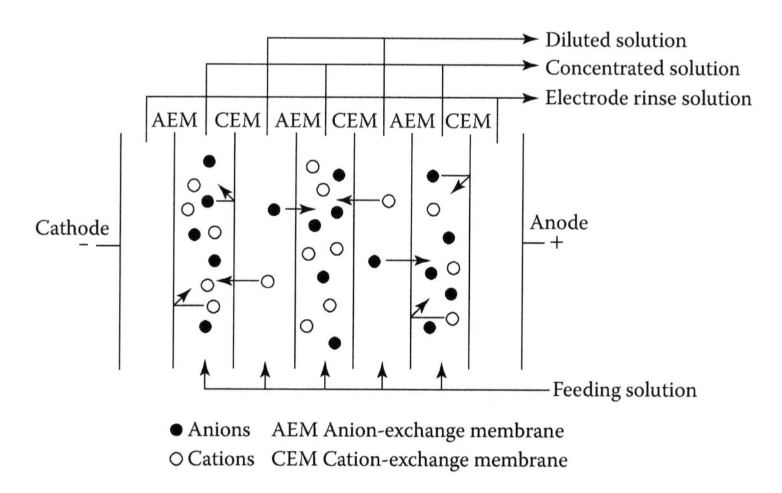

FIGURE 7.5
Schematic diagram of a typical ED Cell. (From Marder, L., Sulzbach, G.O., Bernardes, A.M. and Ferreira, J. Z., *J. Braz. Chem. Soc.*, 14, 610–615, 2003.)

It was reported that an ED cell containing 15 cell pairs of cation- and anion-exchange membranes of 80-cm^2 effective cross-sectional areas was found to be successful in multiple ways. In addition to defluoridation, instantaneous desalination of brackish water (at a reasonable energy cost of 2.5–5 kWh/m^3) of product water was also reported feasible.[91] Membranes used in this cell were prepared from interpolymeric films based on high-density polyethylene (HDPE), linear low-density polyethylene (LLDPE), and styrene-divlnylbenzene, which had an 80:20 ratio of HDPE to LLDPE. The fluoride in the treated water was found to be reduced from 15 to 1.5 mg/L, and total dissolved solid (TDS) in brackish water was found to be reduced from 5000 to 15 mg/L. The ED was found to be successful in removing fluoride from brackish waters of Southern Algeria that had a TDS of 3000 mg/L and a fluoride concentration of 3 mg/L. The ED cell used for this purpose comprised 10 pairs of anion- and cation-exchange membranes (0.18 mm, mono-anion, permselective). However, it was suggested that both RO and ED require a high degree of pre treatment for application. The removal of suspended solids and organics is very much essential to prevent membrane fouling. It is to be noted that proper provision should be made to offset the losses due to the retent formation in the order of 10%–25% of permeate.[1,92] Operations of an ED cell with a capacity of 1.0 m^3/h comprising two anion-exchange membranes using a pre selected membrane in removing fluoride from Moroccan ground water was also reported. The study demonstrated that the desired drinking water quality can be easily obtained by ED under predetermined optimized operating conditions.[93] The recovery rates in ED were higher than in other membrane processes such as RO. The ED can be made operational over longer periods in response to fluctuations in feed concentrations. This makes EC suitable for seasonal operations.[93] Although ED is found to be more economical, the quality of the treated water will be slightly inferior to that of RO. Research suggests that ED is efficient in removing inorganic anions such as fluoride from drinking water.

7.7 Defluoridation Techniques: A Summary

Although many technologies are available for defluoridation, as discussed earlier, with each having its own advantages and limitations, a lasting solution is still at large. Coagulation methods are proved to be unsuccessful in bringing fluoride to desired concentration levels. Sorption methods (including ion-exchange and adsorption techniques) are more effective in reducing fluoride concentrations below permissible limits compared with precipitation methods. Among the adsorbents used, AA was reported successful at the implementation level. Regeneration of used adsorbents and associated treatment costs, reduction in adsorption capacity after regeneration, and pre treatment requirement of pH adjustments could be quoted as the major limitations of most of the adsorbents discussed. Membrane separations are prone to fouling, membrane degradation, and scaling. Further, these are comparatively expensive to install and operate. The electrochemical techniques are constrained by high costs, during both operation and maintenance. The high cost of technology is a limiting factor in the adaptation of treatment systems in many fluoride endemic areas of the developing world. It is important to consider the social acceptance and sustainability of the technology. Needless to say, the "best available technology" need not always be the "most appropriate." Paradoxically, though defluoridation research has made significant advancement, any universal sustainable solution still appears intangible.[1]

7.8 Summary

- The defluoridation techniques generally practiced include coagulation, adsorption/ion exchange, electrochemical methods, and membrane processes.
- Defluoridation processes by coagulation include precipitation of fluoride by a suitable reagent through chemical reactions and/or co-precipitation of fluoride. This involves simultaneous precipitation of fluoride with a macro-component from the same solution through the formation of mixed crystals, by multiple mechanisms such as adsorption, occlusion, or mechanical entrapment.
- Alum treatment is confined to high dosage requirement, issues of sludge disposal, high pH of the treated water, and residual alumina in treated water.
- Adsorption or ion exchange is one of the most frequently used methods for defluoridation. Water laced with fluoride is passed through a column packed with an adsorbent and on saturation, the adsorbent bed is be backwashed for reuse. The adsorption capacity, cost of the adsorbent, ease in operation, and potential for reuse and regeneration are some of the factors that define the selection of an adsorbent.
- The defluoridation potential of activated carbon was found to be poor. The nonmetallic solids such as activated carbons are found to have only weaker attractions for fluoride than that of metallic solids such as activated bauxite or AA.
- AA is a granular porous material with a very high surface area; it consists mainly of aluminum oxide. The application of AA in the defluoridation of drinking water is widely accepted worldwide. The sorption mechanism could be effectively described by the surface

complex formation model depicting the role of surface hydroxo groups in surface complex formation.

- Though many adsorbents are widely used for defluoridation, none of these adsorbents is considered a "universal defluoridation media" given the magnitude and diversity of the problem. The proximity and availability of these adsorbents that are adjacent to a fluorotic endemic area at reasonable or no cost may be the most appropriate choice for consideration.
- The regular replacement of sacrificial electrodes due to continuous dissolution into solution because of oxidation and the high consumption of electric power during operation may also be treated as limitations of the electrocoagulation process.
- The application of membrane technologies in removing inorganic anions such as fluoride is of recent origin. This process turns attractive in water treatment, as many of the difficulties associated with precipitation, coagulation, or adsorption can be either avoided or minimized. Recently, many hybrid systems have been developed in which one membrane process has been integrated with another to produce even higher-quality water.
- The major limitations of RO include the high energy consumption needed for maintaining the required pressure difference. In addition to biological fouling, membranes are subjected to NOM and microorganisms. Mineral fouling may also occur due to precipitation of certain salts. Further, interference from divalent metallic cations such as Ca^{2+} and Mg^{2+} and anions such as SO_4^{2-} may interfere with the separation of particular toxic anionic species. This may affect water recovery and reduce the osmotic pressure demanding pre treatment of raw water.
- The recovery rates in electrodialysis were higher than in other membrane processes such as RO. The ED can be made operational over longer periods in response to fluctuations in feed concentrations, making it suitable for seasonal operations. Though ED is found to be more economical, RO ensures more quality of the treated water compared with it.
- Membrane separations are prone to membrane degradation, scaling, and fouling. It is relatively expensive to install and operate. The electrochemical techniques are constrained by high costs, during both operation and maintenance.
- Although many technologies are available for defluoridation, with each having its own advantages and limitations, a lasting solution is still at large. It appears that "though defluoridation research has made significant advancement, any universal sustainable solution still appears intangible."

References

1. Ayoob, S., Gupta, A.K and Bhat, Venugopal, T. (2008). A conceptual overview on sustainable technologies for the defluoridation of drinking water. *Crit. Rev. Environ. Sci. Technol.*, 38, 6, 401–470.
2. Aksu, Z. and Gönen, F. (2004). Biosorption of phenol by immobilized activated sludge in a continuous packed bed: Prediction of breakthrough curves. *Process Biochem.*, 39, 599–613.
3. Lounici, H., Belhocine, D., Grib, H., Drouiche, M., Pauss, A. and Mameri, N. (2004). Fluoride removal with electro-activated alumina. *Desalination*, 161, 287–293.
4. Hu, C.Y., Lo, S.L. and Kuan, W.H. (2005). Effects of the molar ratio of hydroxide and fluoride to Al(III) on fluoride removal by coagulation and electrocoagulation. *J. Colloid Interface Sci.*, 283, 472–476.
5. Mjengera, H. and Mkongo, G. (2003). Appropriate deflouridation technology for use in flourotic areas in Tanzania. *Phys. Chem. Earth*, 28, 1097–1104.
6. Velizarov, S., Crespo, J.G. and Reis, M.A. (2004). Removal of inorganic anions from drinking water supplies by membrane bio/processes. *Rev. Environ. Sci. Bio/Technol.*, 3, 361–380.
7. Gregory, J. and Duan, J. (2001). Hydrolyzing metal salts as coagulants. *Pure Appl. Chem.*, 73, 2017–2026.
8. Reardon, J.E. and Wang, Y. (2000). A limestone reactor for fluoride removal from wastewaters. *Environ. Sci. Technol.*, 34, 3247–3253.
9. Wasay, S.A., Haran, M.J. and Tokunaga, S. (1996). Adsorption of fluoride, phosphate, and arsenate ions on lanthanum-impregnated silica gel. *Water Environ. Res.*, 68, 295–300.
10. Hu, C.Y., Lo, S.L. and Kuan, W.H. (2003). Effects of co-existing anions on fluoride removal in electrocoagulation (EC) process using aluminum electrodes. *Water Res.*, 37, 4513–4523.
11. Castel, C., Schweizer, M., Simonnot, M.O. and Sardin, M. (2000). Selective removal of Fluoride ions by a two-way ion-exchange cyclic process. *Chem. Eng. Sci.*, 55, 3341–3352.
12. Huang, C.J. and Liu, J.C. (1999). Precipitation flotation of fluoride containing wastewater from semi-conductor manufacture. *Water Res.*, 33, 3403–3412.
13. Lislie, A.L. (1967). Means and methods of defluoridation of water, United States Patent No. 3,337,453.
14. Rao, S.M. and Mamatha, P. (2004). Water quality in sustainable water management. *Curr. Sci.*, 87, 942–947.
15. Dahi, E., Mtalo, F., Njau, B. and Bregnhj, H. (1996). Defluoridation using the Nalgonda technique in Tanzania. In: 22nd WEDC Conference, Reaching the Unreached: Challenges For the 21st Century, New Delhi, India, pp. 266–268.
16. WHO (2006). In: Fawell, J., Bailey, K., Chilton, J., Dahi, E., Fewtrell, L. and Magara, Y. (Eds.), *Fluoride in Drinking Water*, pp. 41–75. London, UK: IWA Publishing.
17. Qureshi, N. and Malmberg, R.H. (1985). Reducing aluminum residuals in finished water. *J. AWWA*, 77, 101–108.
18. Peavy, S.H., Rowe, R.D. and Tchobanoglous, G. (1985). *Environmental Engineering, International Edition*, pp. 134–135. Singapore: McGraw-Hill Book Co.

19. Weber, W.J. Jr. (1972). *Physicochemical Processes for Water Quality Control*, pp. 199–360. New York: A Wiley-Inter science Publication, John Wiley and Sons.
20. Mekonen, A., Kumar, P. and Kumar, A. (2001). Integrated biological and physicochemical treatment process for nitrate and fluoride removal. *Water Res.*, 35, 3127–3136.
21. Hao, J.O. and Huang, C.P. (1986). Adsorption characteristics of fluoride onto Hydrous Alumina. *J. Environ. Eng. (ASCE)*, 112, 1054–1067.
22. Bulusu, K.R. (1984). Defluoridation of water using combination of aluminum chloride and aluminum sulfate. *J. Inst. Eng. (India)*, 65, 22–26.
23. Gupta, S.K., Gupta, A.B., Dhindsa, S.S., Seth, A.K., Agrawal, K.C. and Gupta, R.C. (1999). Performance of domestic filter based on KRASS defluoridation process. *J. Indian Water Works Assoc.*, 31, 193–200.
24. Sujana, M.G., Thakur, R.S. and Rao, S.B. (1998). Removal of fluoride from aqueous solutions by using alum sludge. *J. Colloid Interface Sci.*, 206, 94–101.
25. Choi, W.W. and Chen, K.Y. (1979). The removal of fluoride from waters by adsorption. *J. AWWA*, 71, 562–570.
26. Nawlakhe, W.G., Kulkarni, D.N., Pathak, B.N. and Bulusu, K.R. (1975). Defluoridation of water by nalgonda technique. *Indian J. Environ. Health*, 17, 26–65.
27. CPHEEO (1991). *Central Public Health and Environmental Engineering Organization, Manual on Water Supply and Treatment*. 3rd edn, pp. 289–297. New Delhi, India: The Controller of Publications.
28. RGNDWM (2001). Rajiv Gandhi National Drinking Water Mission, Making water safe; two user friendly methods to deal with the scourage of fluorosis, Jalavani-news letter on rural water and sanitation in India, Published by RGNDWM and Water and Sanitation Program–South Asia (WSP-SA), 3, p.7.
29. Susheela, A.K. (2003). *A Treatise on Fluorosis*, revised 2nd edn. New Delhi, India: Fluorosis Research and Rural Development Foundation.
30. Abe, I., Iwasaki, S., Tokimoto, T., Kawasaki, N., Nakamura, T. and Tanada, S. (2004). Adsorption of fluoride ions onto carbonaceous materials. *J. Colloid Interface Sci.*, 275, 35–39.
31. Bhargava, D.S. and Killedar, D.J. (1991). Batch studies of water defluoridation using fishbone charcoal. *J. Water Pollut. Control Fed.*, 63, 848–858.
32. Zevenbergen, C., van Reeuwijk, L.P., Frapporti, G., Louws, R.J. and Schuiling, R.D. (1996). Simple method for defluoridation of drinking water at village level by adsorption on Ando soil in Kenya. *Sci. Total Environ.*, 188, 225–232.
33. Agarwal, M., Rai, K., Shrivastav, R. and Dass, S. (2003). Deflouridation of water using amended clay. *J. Cleaner Prod.*, 11, 439–444.
34. Zhuang, J. and Yu, G.R. (2002). Effects of surface coatings on electrochemical properties and contaminant sorption of clay minerals. *Chemosphere*, 49, 618–629.
35. Kau, P.M.H., Smith, D.W. and Binning, P. (1998). Experimental sorption of fluoride by kaolinite and bentonite. *Geoderma*, 84, 89–108.
36. Hingston, F.J., Posner, A.M. and Quirk, J.P. (1972). Anion adsorption by goethite and gibbsite I, The role of the protein in determining adsorption envelops. *J. Soil Sci.*, 23, 177–192.
37. Davis, J.A. and Kent, D.B. (1990). Surface complexation modeling in aqueous geochemistry. *Reviews in Mineralogy*, Vol. 23, pp. 177–260. *Mineral-Water Interface Geochemistry*. In: Hochella Jr., M.F., White, A.F. (Eds.), Washington, DC: Mineralogical Society of America.

38. Hiemstra, T. and Van Riemsdijk, W.H. (2000). Fluoride adsorption on Goethite in relation to different types of surface sites. *J. Colloid Interface Sci.*, 225, 94–104.
39. Fan, X., Parker, D.J. and Smith, M.D. (2003). Adsorption kinetics of fluoride on low cost materials. *Water Res.*, 37, 4929–4937.
40. Coetzee, P.P., Coetzee, L.L., Puka, R. and Mubenga, S. (2003). Characterisation of selected South African clays for defluoridation of natural waters. *Water SA*, 29, 331–338.
41. Harrington, L.F., Cooper, E.M. and Vasudevan, D. (2003). Fluoride sorption and associated aluminum release in variable charge soils. *J. Colloid Interface Sci.*, 267, 302–313.
42. Padmasiri, J.P. and Dissanayake, C.B. (1995). A simple defluoridator for removing excess fluorides from fluoride-rich drinking water. *Int. J. Environ. Health Res.*, 5, 153–160.
43. Ramos, R.L., Ovalle-Turrubiartes, J. and Sanchez-Castillo, M.A. (1999). Adsorption of fluoride from aqueous solution on aluminum impregnated carbon. *Carbon*, 37, 609–617.
44. Sinha, S., Pandey, K., Mohan, D. and Singh, K.P. (2003). Removal of fluoride from aqueous solutions by *Eichhornia crassipes* biomass and its carbonized form. *Ind. Eng. Chem. Res.*, 42, 6911–6918.
45. Mohan, S.V., Ramanaiah, S.V., Rajkumar, B. and Sarma, P.N. (2007). Biosorption of fluoride from aqueous phase onto algal *Spirogyra* IO1 and evaluation of adsorption kinetics. *Bioresour. Technol.*, 98, 1006–1011.
46. Dey, S., Goswami, S. and Ghosh, C.U. (2004). Hydrous Ferric Oxide (HFO)—A Scavenger for Fluoride from Contaminated Water. *Water Air Soil Pollut.*, 158, 311–323.
47. Kamble, S.P., Jagtap, S., Labhsetwar, N.K., Thakare, D., Godfrey, S., Devotta, S. and Rayalu, S.S. (2007). Defluoridation of drinking water using chitin, chitosan and lanthanum-modified chitosan. *Chem. Eng. J.*, 129, 173–180.
48. Ma, W., Ya, F.Q., Han, M. and Wang, R. (2007). Characteristics of equilibrium, kinetics studies for adsorption of fluoride on magnetic-chitosan particle. *J. Hazard. Mater.*, 143, 296–302.
49. Sarkar, M., Banerjee, A., Pramanick, P.P. and Sarkar, A.R. (2007). Design and operation of fixed bed laterite column for the removal of fluoride from water. *Chem. Eng. J.*, 131, 329–335.
50. Lv, L., He, J., Wei, M., Evans, D.G. and Zhou, Z. (2007). Treatment of high fluoride concentration water by MgAl-CO_3 layered double hydroxides: Kinetic and equilibrium studies. *Water Res.*, 41, 1534–1542.
51. Kasprzyk-Hordern, B. (2004). Chemistry of alumina, reactions in aqueous solutions and its application in water treatment. *Adv. Colloid Interface Sci.*, 110, 19–48.
52. Nawrocki, J., Dunlap, C., McCormick, A. and Carr, P.W. (2004). Part I. Chromatography using ultra-stable metal oxide-based stationary phases for HPLC. *J. Chromatogr. A*, 1028, 1–30.
53. Knozinger, H. and Ratnasamy, P. (1978). Catalytic aluminas: Surface models and characterization of the surface sites. *Catal. Rev. Sci. Eng.*, 17, 31–70.
54. Nawrocki, J., Rigney, M.P., McCormick, A. and Carr, P.W. (1993). Chemistry of zirconia and its use in chromatography. *J. Chromatogr. A*, 657, 229–282.
55. Blackwell, J.A. and Carr, P.W. (1991). Study of the fluoride adsorption characteristics of porous microparticulate zirconium oxide. *J. Chromatogr.*, 549, 43–57.

56. Daw, R.K. (2004). Experiences with domestic defluoridation in India. In: Sam, G. (Ed.), *People-Centred Approaches to Water and Environmental Sanitation*. Proceedings of the 30th WEDC International Conference, October 2004, pp. 467–473. Vientiane, Lao PDR: Lao National Cultural Hall.
57. Ghorai, S. and Pant, K.K. (2005). Equilibrium, kinetics and breakthrough studies for adsorption of fluoride on activated alumina. *Sep. Pur. Technol.*, 42, 265–271.
58. Chauhan, V.S., Dwivedi, P.K., Iyengar, L. (2007). Investigations on activated alumina based domestic defluoridation units. *J. Hazard. Mater.*, 139, 103–107.
59. Nakkeeran, E. and Sitaramamurthy, D.V. (2007). Removal of fluoride from ground water. *Can. J. Pure Appl. Sci.*, 1, 79–82.
60. Bishop, P.L. and Sansoucy, G. (1978). Fluoride removal from drinking water by fluidized activated alumina adsorption. *J. AWWA*, 70, 554–559.
61. Rubel, F. and Woosley, R.D. (1979). The removal of excess fluoride form drinking water by activated alumina. *J. AWWA*, 71, 45–49.
62. Schoeman, J.J. and MacLeod, H. (1987). The effect of particle size and interfering ions on fluoride removal by activated alumina. *Water SA*, 13, 229–234.
63. Dahi, E. (2000). The State of Art of Small Community Defluoridation of Drinking Water. In: Dahi, E., Rajchagool, S., Osiriphan, N. (Eds.), Proceedings of the 3rd International Workshop on Fluorosis Prevention and Defluoridation of Drinking Water, Chiang Mai, Thailand, pp. 137–167. http://www.icoh.org /download/3rdproceeding.pdf.
64. Cengeloglu, Y., Kir, E. and Ersoz, M. (2002). Removal of fluoride from aqueous solution by using red mud. *Sep. Purif. Technol.*, 28, 81–86.
65. Mohapatra, D., Mishra, D., Mishra, S.P., Chaudhury, G.R. and Das, R.P. (2004). Use of oxide minerals to abate fluoride from water. *J. Colloid Interface Sci.*, 275, 355–359.
66. Das, N., Pattanaik, P. and Das, R. (2005). Defluoridation of drinking water using activated titanium rich bauxite. *J. Colloid Interface Sci.*, 292, 1–10.
67. Li, H.Y., Wang, S., Cao, A., Zhao, D., Zhang, X., Xu, C., Luan, Z., Ruan, D., Liang, J., Wu, D. and Wei, B. (2001). Adsorption of fluoride from water by amorphous alumina supported on carbon nanotubes. *Chem. Phys. Lett.*, 350, 412–416.
68. Zhou, Y., Yu, C. and Shan, Y. (2004). Adsorption of fluoride from aqueous solution on La^{3+} impregnated cross-linked gelatin. *Sep. Purif. Technol.*, 36, 89–94.
69. Olde Damink, L.H.H., Dijkstra, P.J., Van Luyn, M.J.A., Van Wachem, P.B., Nieuwenhuis, P. and Feijen, J. (1995). Glutaraldehyde as a crosslinking agent for collagen based biomaterials. *J. Mater. Sci. Mater. Med.*, 6, 460–472.
70. Xiaoyun, L., Kuanxiu, S., Xiuru,Y. et al. (1999). Development of deflourination from water by rare earth compound. *Chem. Ind. Eng.*, 16, 286–291, as cited in: Zhou, Y., Yu, C., Shan, Y. (2004) Adsorption of fluoride from aqueous solution on La^{3+} impregnated cross-linked gelatin. *Sep. Purif. Technol.*, 36, 89–94.
71. Xiaoyun, L., Jianping, W., Kuanxiu, S. et al. (2001). Properties of resin adsorbent loaded Ce(IV) Ion for removing fluoride ions. *Ion Exchange Adsorption*, 17, 131–137, as cited in: Zhou, Y., Yu, C., Shan, Y. (2004) Adsorption of fluoride from aqueous solution on La^{3+} impregnated cross-linked gelatin. *Sep. Purif. Technol.*, 36, 89–94.
72. Onyango, M.S., Kojima, Y., Aoyi, O., Bernardo, E.C. and Matsuda, H. (2004). Adsorption equilibrium modeling and solution chemistry dependence of fluoride removal from water by trivalent-cation-exchanged zeolite F-9. *J. Colloid Interface Sci.*, 279, 341–350.

73. Tripathy, S.S., Bersillon, J.L. and Gopal, K. (2006). Removal of fluoride from drinking water by adsorption onto alum-impregnated activated alumina. *Sep. Pur. Technol.*, 50, 310–317.
74. Veressinina, Y., Trapido, M., Ahelik, V. and Munter, R. (2001). Fluoride in drinking water: The problem and its possible solutions. *Proc. Estonian Acad. Sci. Chem.*, 50, 81–88.
75. Mameri, N., Yeddou, A.R., Lounici, H., Belhocine, D., Grib, H. and Bariou, B. (1998). Defluoridation of septentrional Sahara water of North Africa by electrocoagulation process using bipolar aluminum electrodes. *Water Res.*, 32, 1604–1612.
76. Ming, L., Yi, S.R., Hua, Z.J., Yuan, B., Lei, W., Ping, L. and Fuwa, K.C. (1983). Elimination of excess fluoride in potable water with coacervation by electrolysis using aluminum anode. *Fluoride*, 20, 54–63.
77. Zhu, J., Zhao, H. and Ni, J. (2007). Fluoride distribution in electrocoagulation defluoridation process. *Sep. Pur. Technol.*, 56, 184–191.
78. Lounici, H., Adour, L., Belhocine, D., Elmidaoui, A., Bariou, B. and Mameri, N. (2001). Novel technique to regenerate activated alumina bed saturated by fluoride ions. *Chem. Eng. J.*, 81, 153–160.
79. Ulbricht, M. (2006). Advanced functional polymer membranes. *Polymer*, 47, 2217–2262.
80. Ndiaye, P.I., Moulln, P., Dominguez, L., Millet, J.C. and Charbit, F. (2005). Removal of fluoride from electronic industrial effluent by RO membrane separation. *Desalination*, 173, 25–32.
81. Schneiter, R.W. and Middtebrooks, E.J. (1983). Arsenic and fluoride removal from ground water by reverse osmosis. *Environ. Int.*, 9, 289–292.
82. Cohen, D. and Conrad, H.M. (1998). 65,000 GPD fluoride removal membrane system in Lakeland, California, USA. *Desalination*, 117, 19–35.
83. Ritchie, S.M.C. and Bhattacharyya, D. (2002). Membrane-based hybrid processes for high water recovery and selective inorganic pollutant separation. *J. Hazard. Mater.*, 92, 21–32.
84. Arora, M., Maheshwari, R.C., Jain, S.K. and Gupta, A. (2004). Use of membrane technology for potable water production. *Desalination*, 170, 105–112.
85. Hu, K. and Dickson, J.M. (2006). Nanofiltration membrane performance on fluoride removal from water, *J. Membr. Sci.*, 279, 529–538.
86. Nicoll, H. (2001). Nanofiltration makes surface water drinkable. *Filtr. Sep.*, 38, 22–23.
87. Bruggen, B.V. and Vandecasteele, C. (2003). Removal of pollutants from surface water and ground water by nanofiltration: Overview of possible applications in the drinking water industry. *Environ. Pollut.*, 122, 435–445.
88. Hagmeyer, G. and Gimbel, R. (1998). Modelling the salt rejection of nanofiltration membranes for ternary ion mixtures and for single salts at di⊠erent pH values. *Desalination*, 117, 247–256.
89. Kettunen, R. and Keskitalo, P. (2000). Combination of membrane technology and limestone filtration to control drinking water quality. *Desalination*, 131, 271—283.
90. Marder, L., Sulzbach, G.O., Bernardes, A.M. and Ferreira, J.Z. (2003). Removal of cadmium and cyanide from aqueous solutions through electrodialysis. *J. Braz. Chem. Soc.*, 14, 610–615.

91. Adhikary, S.K., Tipnis, U.K., Harkare, W.P. and Govindan, K.P. (1989). Defluoridation during desalination of brackish water by electrodialysis. *Desalination*, 71, 301–312.
92. Amor, Z., Malki, S., Taky, M., Bariou, B., Mameri, N. and Elmidaoui, A. (1998). Optimization of fluoride removal from brackish water by electrodialysis. *Desalination*, 120, 263–271.
93. Tahaikt, M., Achary, I., Sahli, M.A.M., Amor, Z., Taky, M., Alami, A., Boughriba, A., Hafsi, M. and Elmidaoui, A. (2006). Defluoridation of Moroccan ground water by electrodialysis: Continuous operation. *Desalination*, 189, 215–220.

8

Adsorptive Removal of Fluoride: A Case Study

8.1 Introduction

This chapter describes a case study dealing with the application of a novel adsorbent, alumina cement granules (ALC), in removing fluoride from groundwater. The sorption capacity of ALC in fluoride uptake was evaluated by various laboratory experiments. The feasibility of its use was first examined by continuously mixed batch reactor (CMBR) studies, commonly referred to as *batch studies*. The kinetics of sorption, equilibrium sorption capacity, and the mechanisms of fluoride removal were mainly evaluated through batch performances. The field application potential of ALC for domestic and community uses was evaluated in terms of its sorptive responses from continuous flow, fixed bed studies, commonly referred to as *column studies.*

8.2 Materials and Methods

8.2.1 Reagents and Adsorbate

All chemicals and reagents used in this study were of analytical grade. NaF (Merck) was used for the preparation of standard fluoride stock solution in double-distilled water. All synthetic fluoride solutions for adsorption and analysis were prepared by an appropriate dilution of the stock solution in de ionized (DI) water. The natural fluoride-rich drinking water was collected from Baliasingh Patna, a fluoride endemic village (Kurda district, Orissa state) in India. Only plasticware was used for handling fluoride solution, and it is neither prepared in nor added to glass containers. All plasticware and glassware were pre soaked in a dilute HNO_3 acid bath, washed with dilute soap solution, rinsed thoroughly with DI water and dried prior to use.[1]

8.2.2 Synthesis of the Adsorbent

The adsorbent ALC, selected for the present research, was prepared from commercially available high alumina cement. The high proportion of alumina and calcium, whose potential in fluoride removal was established, was instrumental in selection. Initially, slurry was prepared by adding distilled water to 1 kg of high alumina cement at a water–cement ratio of 0.3. The slurry was kept at an ambient temperature for 2 days for setting, drying and hardening. This hardened paste was cured in water for 5 days. After curing, it was broken, granulated, sieved to a geometric mean size of around 0.212 mm, and kept in airtight containers for use.[1,2]

8.2.3 Instrumentation

The elemental composition of ALC was determined by energy-dispersive x-ray (EDX) analysis (Oxford ISIS-300 model) by the quantitative method in two iterations by using ZAF correction, at a system resolution of 65 eV, and results were normalized stoichiometrically. The surface area of the adsorbent was determined by the Brunauer, Emmett and Teller (BET) method at liquid nitrogen temperature by using FlowSorb II 2300 (Micrometrics Instruments corporation, USA). The chemical composition of ALC was determined by x-ray diffraction analysis (XRD) by using a Miniflex diffractometer (30 kV, 10 Maq; Rigaku Corp., Tokyo, Japan) with a Cu Kα source and a scan rate of 2°/min at room temperature. The Nexus™ 870 spectrometer (Thermo Nicolet) was used for Fourier transform infrared (FTIR) analysis. Expandable Ion Analyzer EA 940 with Orion ion plus (96-09) fluoride electrode (Thermo Electron Corporation, Beverly, Massachusetts), using TISAB III buffer, was used for fluoride measurement. The pH measurement was done by a Cyber Scan 510 pH meter (Oakton Instruments, USA). A temperature-controlled orbital shaker (Remi Instruments Ltd., Mumbai, India) was used for agitation of the samples in batch studies. A high-precision electrical balance (Mettler Toledo, Model AG135) was used for the weight measurement. A flame atomic absorption spectrophotometer (FAAS) (Shimadzu, Model AA–6650) was used for quantitative analysis of elements such as iron, calcium and magnesium. Conductivity and total dissolved solids (TDS) were measured using Cyber Scan 510, Eutech instruments, Singapore. Nephelo turbidity meter (Systronics, Model 131) was used for the measurement of turbidity.[2–4]

Peristaltic pumps (Miclins, Chennai, India) were used for controlling flow rates in column studies.[5]

8.2.4 Characterization of the Adsorbent

The developed adsorbent ALC was characterized by EDX, scanning electron microscopy (SEM), XRD, FTIR, and various physicochemical analyses. EDX analysis was used to determine the elemental composition of ALC (combined with oxygen). SEM photographs of the adsorbent were taken to study the surface texture of ALC grains. The XRD analyses were carried out to identify the morphological structure and the extent of crystallinity of the adsorbent. The FTIR analyses were done to understand the spectroscopic features of the adsorbent. The bulk density was determined by pouring 5 g of ALC into a 100-mL-stoppered measuring cylinder half filled with water. This was thoroughly mixed by inverting the stoppered measuring cylinder several times. The adsorbent was then allowed to settle to a constant volume. The bulk density is reported in g/cm^3. The pH at zero-point charge of ALC was also determined. Different quantities of ALC were placed in 10-mL solutions of 0.1 M NaCl (prepared in pre boiled water) in various bottles and kept in a thermostat shaker for overnight continuous agitation. The equilibrium pH values of these mixtures were measured, and the limiting value was reported as pH_{zpc}.[3] The other chemical analyses were conducted as per APHA guidelines.[6]

8.3 Batch Studies

Batch studies were performed to investigate the sorptive characteristics of the developed adsorbent ALC. Thus, experiments were conducted on the adsorption and desorption of fluoride in single-component aqueous systems, unless otherwise specified. The experimental studies include kinetics, equilibrium, thermodynamic profiles of sorption, and influence of various parameters on the sorption of fluoride onto ALC. The batch sorption experiments were conducted using polyethylene bottles (Tarson Co. Ltd., India) of 150-mL capacity with 50 mL of fluoride solutions of a desired concentration and pH. No pH adjustments were made unless otherwise mentioned. ALC was added as per dose requirements; bottles were capped tightly and shaken in the orbital shaking incubator at 230 ± 10 rpm at room temperature (300 ± 2 K), unless otherwise mentioned.

> The bottles were taken out from the shaker at the desired time interval and filtered using Whatman No-42 filter paper to separate the sorbent and filtrate. From the filtered sample of each batch reactor, 10 mL was taken for the analysis and determination of residual fluoride. All batch sorption experiments were duplicated with an experimental error limit ±5%, and average values were reported. In order to check for any adsorption on the walls of the container, blank container adsorption tests were also carried out.[2,3]

8.3.1 Effect of Process Parameters

Experiments to study the effects of agitation rate on fluoride removal were conducted at a fixed dose of ALC (1.5 g/L in synthetic water and 8 g/L in natural water) at an initial pH of 6.9 ± 0.4 and initial fluoride concentrations of 8.65 mg/L. The samples were shaken at varying agitation rates of

120–280 rpm for 4 h at a temperature of 300 ± 2 K. Experiments to study the effects of adsorbent dose on the removal of fluoride were conducted by varying the dose of ALC (0.25–3.0 g/L in synthetic water and 3–12 g/L in natural water) at an initial pH of 6.9 ± 0.4 with an initial fluoride concentration of 8.65 mg/L. The samples were shaken at an agitation rate of 230 ± 10 rpm for 3 h at a temperature of 300 ± 2 K. Experiments to study the effects of contact time on fluoride removal were conducted at a fixed dose of ALC (2 g/L in synthetic water and 10 g/L in natural water) at an initial pH of 6.9 ± 0.4 and initial fluoride concentrations of 8.65 mg/L (agitation rate of 230 ± 10 at 300 ± 2 K). The samples were taken at the first 2, 4, 8, 10, 30, 45, 60, 75, 90, 150, 180, 210, and 240 min and analyzed for remaining fluoride concentrations. Experiments to study the effects of initial adsorbate concentration were conducted at a fixed dose of ALC (2 g/L in synthetic water) at an initial pH of 6.9 ± 0.4. Initial fluoride concentrations of 2.5, 5, 8.65, 25, and 50 mg/L were taken and agitated at 230 ± 10 rpm at 300 ± 2 K. The samples were taken at the first 10, 30, 45, 60, 90, 120, 150, 180, 210, and 240 minutes and analyzed for remaining fluoride concentrations.

8.3.2 Equilibrium Studies

The equilibrium sorption studies were conducted in natural water with varying amounts of ALC in the range of 3–12 g/L at its natural concentration of 8.65 mg/L and pH of 6.9 ± 0.4 at 230 ± 10 rpm for 3 h (equilibrium time). In synthetic water, equilibrium sorption studies were conducted by both variations in doses of ALC (dose variation study) and variations in initial fluoride concentrations (concentration variation study). In the dose variation study, the dose range of ALC selected was 0.25–3.0 g/L at an initial fluoride concentration of 8.65 mg/L (pH = 6.9 ± 0.4, agitation rate = 230 ± 10 rpm, equilibrium agitation time = 3 h). In concentration variation studies in synthetic water, the range of initial fluoride concentration selected was 2.5–100 mg/L at a fixed ALC dose of 2 g/L (pH = 6.9 ± 0.4, agitation rate = 230 ± 10 rpm, equilibrium agitation time = 3 h). The effect of initial pH on the removal of fluoride in synthetic water was studied over a lower ALC dose of 1.5 g/L for the initial fluoride concentration of 8.65 mg/L (agitated at 230 ± 10 rpm for 3 h at 300 ± 2 K). Then, 2M HCl and 2M NaOH solutions were used to adjust the pH of the synthetic solution. Sorption experiments were conducted at 290, 300, and 310 K to study the effects of temperature on the removal of fluoride. The experiments were conducted on both synthetic and natural water at a fixed ALC dose (2 g/L in synthetic water and 10 g/L in natural water) and an initial fluoride concentration (8.65 mg/L). The agitation time and rate provided were 3 h and 230 ± 10 rpm, respectively, at all temperatures of the study. To study the effects of ionic strength or inert electrolyte concentrations on the sorption of fluoride onto ALC, sodium nitrate solutions of strength ranging from 10^{-4} to 10^{-1} M were used in synthetic water. The dose of ALC was 1.5 g/L for the initial fluoride concentration of

8.65 mg/L (agitation rate = 230 ± 10 rpm, equilibrium agitation time = 3 h, temperature = 300 ± 2 K). Batch equilibrium experiments were conducted in synthetic water to elucidate the individual effects of some common ions such as nitrate, chloride, sulfate, bicarbonate, silicate, calcium, iron, and phosphate on the sorption of fluoride onto ALC. The effects of humic acid on fluoride sorption were also investigated. The interference effects of other ions were conducted in synthetic samples having fluoride concentrations of 8.65 mg/L at an ALC dose of 1.5 g/L (agitation rate = 230 ± 10 rpm, equilibrium agitation time = 3 h, temperature = 300 ± 2 K). The salts $NaNO_3$, KCl, Na_2SO_4, $NaHCO_3$, Na_2SiO_3, $Ca(NO_3)_2$, $Fe(NO_3)_2$, and K_2HPO_4 (Merck) were used for preparing the respective ions for interference studies. Humic acid solutions were prepared from humic acid sodium salt. Batch desorption studies were carried out by the dilution method[7] using the spent ALC that had been previously used for fluoride sorption. After completing adsorption up to the equilibrium time of 3 h, 20 mL of the supernatant in the bottle was replaced by an equal volume of 10% NaOH. The batch desorption experiments were conducted as per the same experimental procedure followed in sorption studies.

8.3.3 Column Studies

Continuous flow, fixed bed column experiments were conducted using glass columns of a length of 550 mm and an internal diameter of 20 mm. The column was packed with a desired depth of ALC between two layers of glass wool at the top and bottom ends to prevent the absorbent from floating. The column was fed continuously with the feed solution (natural groundwater/synthetic fluoride water) having the desired concentration at the desired volumetric flow rate by using peristaltic pumps (Miclins, India) in the up-flow mode. The schematic diagram of the column setup is shown in Figure 8.1. The effluent samples were collected at pre determined time intervals and analyzed for the remaining fluoride concentration. The breakpoint concentration of fluoride was taken as 1.00 mg/L according to the permissible limit in drinking water set by the World Health Organization (WHO)[8,9] standards. The exhaust concentration of fluoride was considered 90% of the influent (i.e., $0.9C_0$). All studies were performed at a constant temperature of 300 K to be representative of the prevailing environmental conditions. No pH adjustments were made in natural water. The pH of synthetic water was maintained similar to that of natural water. The effects of various process parameters such as bed depth and flow rate on the sorption profile of ALC were investigated in both natural and synthetic waters. The effects of initial fluoride concentrations in synthetic water were also investigated. The effects of bed depth on the sorption profile of ALC in both synthetic and natural water were investigated by varying the bed depths (5–15 cm) while keeping the influent feed (fluoride) concentration (8.65 mg/L) and influent flow rate (8 mL/min) fixed.

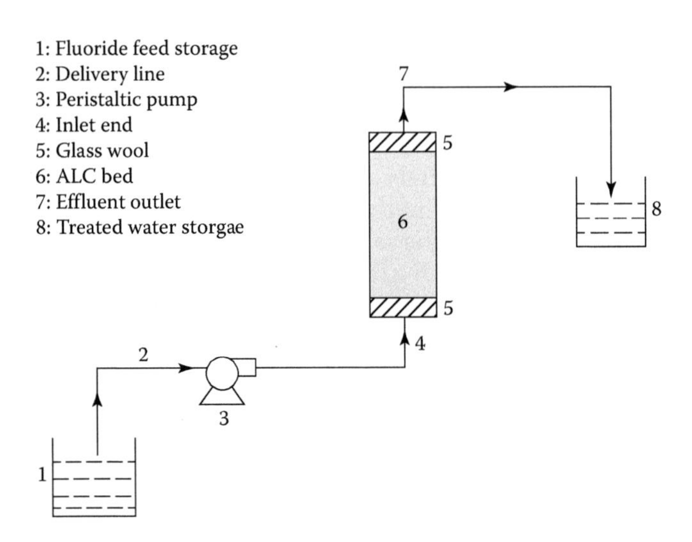

FIGURE 8.1
Experimental setup for fixed bed column studies. (Modified from Ayoob, S., Gupta, A.K. and Bhakat, P.B., *Sep. Purif. Technol.*, 52, 430–438, 2007.)

The effects of feed flow rates on the sorption profile of ALC in both synthetic and natural water were investigated by varying the flow rates (4–12 mL/min) while keeping the influent feed (fluoride) concentration (8.65 mg/L) and bed depth (10 cm) fixed. The effects of initial sorbate concentrations on the sorption profile of ALC in synthetic water were investigated by varying the initial sorbate (fluoride) concentrations (4–15 mg/L) while keeping the flow rate (8 mL/min) and bed depth (10 cm) fixed. The exhausted ALC fixed beds after lengthy column runs (bed depth = 5 cm; flow rate = 8 mL/min; initial fluoride concentration = 8.65 mg/L) of both natural and synthetic waters were regenerated using sodium hydroxide (10%, w/v) solution. The desorption studies in columns were conducted in the same manner as sorption studies, replacing the feed solution with 10% NaOH solutions at a very low flow rate of 0.5 mL/min.

8.4 Theoretical and Mathematical Formulations

8.4.1 Adsorption Capacity

The amount of fluoride adsorbed per unit mass of ALC at any time t (q_t, mg/g) and the adsorption efficiency (%R, determined as the fluoride removal percentage relative to the initial concentration) of the system were calculated as follows:[3]

$$q_t = \frac{C_0 - C_t}{m} V \tag{8.1}$$

$$\%R = \frac{C_0 - C_t}{C_0} \times 100 \tag{8.2}$$

where C_0 and C_t are the fluoride concentrations in solution (mg/L) initially and at any time (t), respectively, m is the mass of ALC (g), and V is the volume in liters of the solution. In Equation 8.1, when $C_t = C_e$ (fluoride concentrations remaining in solution at equilibrium in mg/L), $q_t = q_e$ (equilibrium adsorption capacity in mg/g).

8.4.2 Kinetic Modeling

Prediction of the rate at which adsorption takes place for a given system is probably the most important factor in adsorption system design, as the adsorbate residence time and reactor dimensions are controlled by the kinetics of the system. A number of adsorption processes for pollutants have been studied in an attempt to find a suitable explanation for the mechanisms and kinetics for sorting out environment solutions.[11] Based on kinetic data, various models have been suggested that throw light on the mechanisms of adsorption and potential rate-controlling steps such as mass transport and chemical reaction processes.[12] Kinetic models based on the capacity of the adsorbents mainly include the Lagergren's pseudo-first-order equation and pseudo-second-order equation models.[11,13]

8.4.2.1 Pseudo-First-Order Model

In 1898, Lagergren presented the first-order rate equation for the adsorption of solutes from a liquid solution onto charcoal.[13] Lagergren's equation was the first kinetic equation for the sorption of the liquid/solid system based on solid capacity, and it was used extensively to describe the sorption kinetics. This model assumes that the rate of change of solute uptake with time is directly proportional to the difference in saturation concentration and the amount of solid uptake with time. In order to distinguish a kinetic equation based on the adsorption capacity of a solid from one based on the concentration of a solution, Lagergren's first-order rate equation has been called a *pseudo-first-order equation* and is expressed as follows:

$$\frac{dq_t}{dt} = k_{s1}(q_e - q_t) \tag{8.3}$$

Integration within the boundary conditions $t = 0$ to $t = t$ and $q_t = 0$ to $q_t = q_e$ gives the linearized form as follows:

$$\ln(q_e - q_t) = \ln q_e - k_{s1}t \tag{8.4}$$

where q_e is the amount of soluted sorbate sorbed at equilibrium (mg/g), q_t is the amount of soluted sorbate on the surface of the sorbent at any time t (mg/g), and k_{s1} is the pseudo-first-order rate constant (min^{-1}). The plot of $\ln(q_e - q_t)$ versus t gives a straight line for first-order kinetics, which allows computation of the adsorption rate constant, k_{s1}. When adsorption is preceded by diffusion through a boundary, the kinetics in most cases follows Lagergren's pseudo-first-order rate equation. If the experimental results do not follow Equations 8.3 and 8.4, the sorption is not diffusion controlled and the data differ in two important aspects: (1) $k_{s1}(q_e - q_t)$ does not represent the number of available sites and (2) $\ln q_e$ is not equal to the intercept of the plot of $\ln(q_e - q_t)$ against t.[12,14,15] Then, the reaction is not likely to be first order, irrespective of the magnitude of the correlation coefficient.

8.4.2.2 Pseudo-Second-Order Model

If the rate of sorption is a second-order mechanism, the pseudo-second-order chemisorption kinetic rate equation is expressed as follows:[12]

$$\frac{dq_t}{dt} = k(q_e - q_t)^2 \tag{8.5}$$

Rearranging and integration within the boundary conditions $t = 0$ to $t = t$ and $q_t = 0$ to $q_t = q_e$ give the linearized form as follows:

$$\frac{1}{q_e - q_t} = \frac{1}{q_e} + kt \tag{8.6}$$

which is the integrated rate law for a pseudo-second-order reaction. Rearranging it again, Equation 8.6 is reduced to

$$\frac{t}{q_t} = \frac{1}{h} + \frac{1}{q_e}t \tag{8.7}$$

where k is the pseudo-second-order rate constant (g/mg/min), and h is the initial sorption rate (mg/g/min), which is given by

$$h = kq_e^2 \tag{8.8}$$

The *rate* of a reaction is defined as the change in the concentration of a reactant or product per unit time. Concentrations of products do not appear in the rate law, because the reaction rate is studied under conditions where the reverse reactions do not contribute to the overall rate. The reaction order and rate constant must be determined by experiments. In order to distinguish the kinetics equation based on the concentration of a solution from the adsorption capacity of solids, this second-order-rate equation is called a pseudo-second-order one. The pseudo-second-order rate expression was used to describe chemisorption involving valency forces through the sharing or exchange of electrons between the adsorbent and adsorbate as covalent forces, and ion exchange. The pseudo-second-order model constants can be determined experimentally by plotting t/q_t against t (Equation 8.7). Although there are many factors that influence the adsorption capacity, including initial adsorbate concentration, reaction temperature, solution pH, adsorbent particle size and dose, and nature of the solute, a kinetic model is concerned only with the effect of observable parameters on the overall rate.[11] It was also suggested that the pseudo-second-order equation may be applied to chemisorption processes with a high degree of correlation in several literature cases where a pseudo-first-order rate mechanism has been arbitrarily assumed.[12] Recently, the pseudo-second-order rate equations have been widely applied to the adsorption of pollutants from aqueous solution, since the model has the following advantages: the adsorption capacity, the rate constant, and the initial adsorption rate, all of which can be determined from the equation without knowing any parameter beforehand.[11]

8.4.2.3 Intra Particle Diffusion Model

The concentration dependence of the rate of sorption is commonly used as a partial test of hypothesis regarding the nature of the *rate-controlling step*. Accordingly, most of the recent studies involving adsorption of fluoride on metal oxides explored the use of the intra particle surface diffusion model[16,17] represented by Equation 8.9 to elucidate its mechanism[15,18] as follows:

$$q_t = k_p t^{1/2} + C \tag{8.9}$$

where k_p is the intra particle diffusion rate constant (mg/g/h$^{1/2}$). The value C (mg/g) in this equation is a constant that gives an idea about the thickness of the boundary layer (the larger the value, the greater is the boundary effect). When the plot (q_t vs. $t^{1/2}$) does not pass through the origin, it is indicative of some degree of boundary layer control. This behavior indicates that the intra particle diffusion is not the only rate-limiting step, but other kinetic factors also may control the rate of adsorption, all of which may be operating simultaneously. The slope of linear plot can be used to derive values for the rate parameter, k_p, for the intra particle diffusion. However, if the data exhibit multi linear plots, then two or more steps influence the sorption process.

8.4.2.4 Elovich Equation

The Elovich equation has also been successfully applied in aqueous systems to describe the adsorption and desorption reactions[19,20] as follows:

$$q_t = \frac{1}{\beta}\ln(1 + \alpha\beta t) \tag{8.10}$$

where α and β are constants, t is the time, and q_t is the surface coverage. The Elovich equation can be derived from either a diffusion-controlled process or a reaction-controlled process. If the Elovich equation is based on adsorption on an energetically heterogeneous surface, the parameter β is related to the distribution of activation energies. In the diffusion control model, it is a function of particle and diffusion coefficients. When the term $\alpha\beta t$ is much greater than 1, the equation just cited can be simplified to[19]

$$q_t = \frac{1}{\beta}\ln(\alpha\beta t) = \frac{1}{\beta}\ln(\alpha\beta) + \frac{1}{\beta}\ln(t) \tag{8.11}$$

If the results follow an Elovich equation, the kinetic results will be linear on a q_t versus $\ln t$ plot. It was suggested that the diffusion accounted for the Elovich kinetics pattern[21]; that conformation to this equation alone might be taken as evidence that the rate-determining step is diffusion in nature; and that this equation should apply to conditions where the desorption rate can be neglected.[22]

8.4.2.5 Arrhenius Equation

It was suggested that the value of energy of the activation value (E_a), obtained from Arrhenius equation (Equation 8.12), could also be a useful kinetic parameter in assessing rate-limiting steps.[23] Low E_a values usually indicate diffusion-controlled transport and physical adsorption processes, whereas higher E_a values indicate chemical reactions or surface-controlled processes:

$$K = A_f \exp\left(-\frac{E_a}{RT}\right) \tag{8.12}$$

where K is the rate coefficient of the sorption reaction (g/mg/min), A_f is the pre-exponential factor or frequency factor (g/mg/min), E_a is the activation energy of sorption (kJ/mol), R is the gas constant (8.314 J/mol/K), and T is the thermodynamic temperature (K). Some of the assigned values of E_a (kJ/mol) include 8–25 to physical adsorption, less than 21 to aqueous diffusion, 20–40 to pore diffusion, and greater than 84 to ion exchange.[23] In this study, the kinetic data of fluoride sorption onto ALC were fitted to all the kinetic models cited earlier. The activation energy of the sorption process in both systems was also found out. The experimental investigations as suggested next were carried out to identify the rate-limiting step.

8.4.3 Elucidation of Rate-Limiting Step

> The three important parameters that may determine the kinetics and rate-limiting process of a sorption system are being identified as pH, concentration of inert electrolyte and desorption pattern of the adsorbent.[3,19]

In solutions, the effective particle sizes are influenced by the concentration of inert electrolyte, with larger aggregated particles being formed in the higher concentrations of inert electrolytes at a given pH. Increasing the effective particle size will increase the diffusion path length from the external surface of the aggregate to the reactive sites located on the internal surface of the adsorbent. As a result, it will take a longer time for the ions to reach these reactive sites from an external solution, suggesting the rate-limiting step. It was suggested that the slow stage sorption, indicative of the rate-limiting step, may be due to two reasons: (1) diffusion or (2) surface reactions.[3,19]

As defined earlier, the diffusion may be either inter particle or intra particle. The surface reactions include surface precipitation and surface site bonding energy heterogeneity or other surface reactions.

The following procedure was adopted to elucidate the rate-limiting step:

1. If the inter particle diffusion is rate limiting, the adsorption patterns should be sensitive to inert electrolyte concentrations or pH.
2. If the intra particle diffusion is rate limiting, the desorption pattern should follow the same two-stage pattern as that of sorption.
3. If both inter particle and intra particle diffusion are not rate controlling, by elimination, the surface reactions will be rate limiting.[3,19]

So, in this study, different kinetic models were examined to describe the sorption kinetics; further, the response of the adsorbent to pH and inert electrolyte concentration were also examined to elucidate the rate-limiting mechanism.

8.4.4 Adsorption Equilibrium and Isotherms

In adsorption, in addition to the surface unsaturation of the solid phase, the intermolecular interactions in the liquid phase also play a dominant role. As a result, in general, adsorption is viewed as a complex phenomenon. During the process of adsorption, the molecules of the solute are removed from the solution and transferred onto the adsorbent. This transfer of the solute from the solution to the adsorbent continues until the concentration of the solute remaining in solution is in equilibrium with the concentration of the solute adsorbed by the adsorbent. This state is in a dynamic stability, that is, the amount of solute migrating onto the adsorbent is counterbalanced by the amount of solute migrating back into the liquid phase. The position of equilibrium depends on various parameters such as solute concentration, characteristics of the adsorbent, solvent temperature, and pH. The adsorption isotherms are used for understanding and quantifying this distribution of the adsorbate between the liquid phase and the solid adsorbent phase that are at equilibrium during the adsorption process and also for understanding the mechanism of adsorption. Fitting of experimental data to adsorption isotherm models is an important step in designing and optimizing an adsorption system.[4]

The adsorption isotherm function is an important experimental parameter that measures adsorption as a function of equilibrium solute concentration (C_2), at a given temperature (T). It is expressed quantitatively in terms of the moles of solute species adsorbed per gram of the adsorbent (n_2^s) and can be written as follows:[24]

$$n_2^s = f(C_2, T) \tag{8.13}$$

At constant temperature,

$$n_2^s = f_T(C_2) \tag{8.14}$$

Various functional forms for f have been proposed either by empirical observations or in terms of specific models. The Langmuir, Freundlich, and Dubinin-Radushkevich (D-R) isotherm models are particularly important examples.[25–28]

8.4.4.1 Langmuir Isotherm

Though originally derived for the solid–gas interface, the general kinetic features of the Langmuir model (Equation 8.15) are equally applicable for any interface:

$$q_e = \frac{q_{max} b C_e}{1 + b C_e} \tag{8.15}$$

where q_e is the amount of adsorbate adsorbed at equilibrium per unit weight of adsorbent (mg/g), C_e is the equilibrium solute concentration (mg/L), q_{max} is the saturated monolayer adsorption capacity (mg/g), and b is the Langmuir constant related to the binding energy or affinity parameter (L/mg) of the sorption system, respectively. The corresponding linear form of the equation is as follows:

$$\frac{1}{q_e} = \frac{1}{b q_{max} C_e} + \frac{1}{q_{max}} \tag{8.16}$$

This model assumes a uniform surface with equivalent adsorption sites[29] with no lateral interactions between adsorbed species. The adsorption is limited to a monolayer[24]; adsorption energy is assumed to be independent of the occupancy of neighbouring sites and, hence, has a constant value. There is also no transmigration of the adsorbate in the plane of the surface.

The Langmuir equation includes two constants, each of which has a clear physical meaning: "b" is the equilibrium constant for the adsorption process expressed in terms of the ratio of the adsorption and desorption rate constants and, hence, is directly related to the binding energy ($b\alpha e^{-\Delta H/RT}$, where ΔH is the net enthalpy change).[4] q_{max} is the adsorption limit obtained at high solute concentrations when $bC_e >> 1$, and q_e shows a zero-order dependence on the solute concentration. For very low values of C_e, the term $bC_e << 1$ reduces the hyperbolic equation to a first-order linear equation in solute concentration as follows[4]:

$$q_e = k\,C_e \tag{8.17}$$

where

$$k = q_{max} b \tag{8.18}$$

The affinity parameter (b) is best estimated from the slope of the adsorption isotherm at very low concentrations. However, this slope gives the product $q_{max}\,b$ and not just b. In order to separate these two parameters, it is necessary to know the adsorption maximum q_{max}, and this can only be estimated with precision from data at very high concentrations where the slope of the isotherm approaches zero. If data are restricted to an intermediate range of concentration, they may be fitted very well, but it will be difficult to separate q_{max} and b, the values of which will show a high negative correlation and correspondingly, high standard errors.[4,30]

8.4.4.2 Freundlich Isotherm

Similar to the Langmuir model, the Freundlich model (Equation 8.19) also has an empirical origin, but it is extremely useful for experimentally determining the adsorption capacity (k_f):

$$q_e = k_f C_e^{1/n} \tag{8.19}$$

where n is a constant representing adsorption intensity and is always greater than unity. The corresponding linear form of the equation is as follows:

$$\ln q_e = \ln k_f + \frac{1}{n} \ln C_e \tag{8.20}$$

The Freundlich equation is a special case for heterogeneous surface energies in which the binding energy term b, in the Langmuir equation, varies as a function of the surface coverage q_e, essentially due to variations in heats of adsorption.[24] This model agrees quite well with the Langmuir isotherm at moderate concentrations. However, unlike the Langmuir equation, it is not reduced to a linear adsorption expression at low solute concentrations but remains convex to the concentration axis. It also does not agree with the Langmuir equation at very high solute concentrations, since n must reach a finite limiting value when the surface is fully covered.[4]

The intercept and slope of its linear plot gives a measure of the adsorption capacity (k_f), and an inverse measure of the intensity of adsorption (n), respectively.

8.4.4.3 Dubinin–Radushkevich (D–R) Isotherm

The D–R isotherm (Equation 8.21) is more general than the Langmuir isotherm, since it does not assume a homogeneous surface or constant sorption potential. It was applied to distinguish between the physical and chemical adsorption of metal ions and has limited utility[28]:

$$q_e = Q_m \exp(-k_{ad}\varepsilon^2) \tag{8.21}$$

where k_{ad} is a constant related to adsorption energy, Q_m is the theoretical saturation capacity (mg/g), and ε is the Polanyi potential and is given by

$$= RT \ln\left(1 + \frac{1}{C_e}\right) \tag{8.22}$$

where R is the universal gas constant (8.314×10^{-3} kJ/mol/K) and T is the absolute temperature (K). The corresponding linear form of Equation 8.21 is as follows:

$$\ln q_e = \ln Q_m - k_{ad}\varepsilon^2 \tag{8.23}$$

The nature of interaction between the adsorbate and adsorbent binding sites can be evaluated by the mean free energy of sorption per mole of the adsorbate calculated by the following equation[31]:

$$E = -(2k)^{-0.5} \tag{8.24}$$

where k is the constant obtained from the D–R isotherm. This free energy term is the work done while transferring one mole of the adsorbate to the surface from infinity in solution. If the magnitude of E is between 8 and 16 kJ/mole, the adsorption process proceeds by the exchange mechanism (ion exchange); if it is less than 8 kJ/mole, physisorption occurs.[32]

> The Langmuir and Freundlich isotherm models are said to suffer from two major drawbacks. First, the model parameters obtained are usually appropriate for a particular set of conditions and these cannot be used as a prediction model for another set. Second, these models are unable to provide a fundamental understanding of ion adsorption.[4,33]

Although many other isotherm models have been developed,[34,35] it can be seen that the Langmuir and Freundlich isotherms still remain the two most commonly used equilibrium adsorption equations, due to their simplicity and the ease with which their adjustable parameters can be estimated. This trend generally holds for defluoridation research in which the Langmuir and Freundlich isotherm models have found wide applicability in determining the adsorption capacities of various adsorbents.

8.4.4.4 Selection of Best-Fitting Isotherm

The selection of the *best-fit* isotherm model is a significant step in adsorption studies. Generally, the success of the adsorption process hinges on the development and application of suitable adsorbents. The criteria for selection of an adsorbent mainly include its adsorption capacity (mg of fluoride adsorbed/g of adsorbent) and the cost. The adsorption capacity parameter (q_{max}, k_f, and Q_m by Langmuir, Freundlich, and D–R) can be obtained from the respective isotherm models.

8.4.4.5 Natural and Synthetic Systems

> Generally, the adsorption capacity of an adsorbent is assessed on the basis of equilibrium sorption data that are generated by batch studies through the *best-fitting* of isotherm models. So, the fitting of experimental data to the isotherm models turns out to be the most important, as it dictates the optimum model parameters. However, most of the reported isotherm studies are conducted on concentration variations in synthetic samples (DI or distilled water laced with fluoride), which lose the *real-life* flavors and characteristics of natural samples. In general, apart from concentrations of adsorbate, the pH of the medium and the presence of other competing ions drastically alter the adsorption capacity of the adsorbents in aqueous solutions.[4,36] So, it is rational to suggest that the adsorption capacity reported in synthetic samples cannot be a reliable representation of its actual scavenging potential in field applications.[4]

This observation becomes significant in the context of the reported reduction in the adsorption capacities of activated alumina in removing fluoride in the field studies. Though high fluoride removal capacity of activated alumina (4–15 mg/g) is reported in literature, field experiences demonstrate that it is often around 1 mg/g only.[37] So, in this study, the adsorption capacity of ALC was tested in natural fluoride-rich groundwater. The synthetic water of the same concentration as the natural water (8.65 mg/L) was prepared to have a reliable comparison of the sorption profile with natural water. Further, the effects of various ions (naturally expected in groundwater), pH, ionic strength, and temperature on the sorption behavior of ALC were also investigated.

8.4.4.6 Concentration and Dose Variation Studies

In defluoridation research involving adsorption, it has also become a common practice to compare the adsorption capacities of various adsorbents to highlight their effectiveness and relative trustworthiness. However, of late,

> the limited success of adsorbents in field applications raises apprehensions over the use of adsorption capacity (generated from equilibrium data) as a measure of their effectiveness in drinking water treatment. The batch studies for equilibrium data can be conducted by changing either the concentrations of adsorbate (fluoride) or the dose of adsorbent (ALC). Generally, the adsorptive capacity of the adsorbent will increase with an increase in influent concentrations, till it reaches a maximum. Obviously, the isotherm studies performed for a higher range of fluoride concentrations will show a higher capacity than that at moderate or lower ranges. However, this maximum capacity may not describe the media behavior at typical influent fluoride concentrations. It is reported that the maximum natural fluoride concentrations in the groundwaters of India are in the range of 0.5–48 mg/L. However, it is not uncommon to encounter isotherm studies performed at extremely higher fluoride concentrations. In isotherm studies of natural samples, this approach (of concentration variation studies) becomes inappropriate, as the concentration of fluoride in natural groundwater remains constant. So, in defluoridation studies dealing with natural groundwater, it is only possible to have isotherm studies with dose variations of adsorbents. This further (in general) suggests the irrelevance of concentration variation studies in the adsorption process for removing pollutants such as fluoride from aqueous systems. So, in this study, in addition to concentration variation studies, isotherm studies were also conducted by dose variations of ALC to have a reliable field defluoridating capacity.[4]

8.4.5 Factors Influencing Adsorption

Adsorption is a complex process that must be carefully analyzed in order to ensure optimum process design. Many factors influence adsorption reactions and the rate and extent to which adsorption occurs.

8.4.5.1 Adsorbent Dose

Since each adsorbent particle has to purify a certain volume of liquid, at a constant solute concentration, higher adsorbent dosages will be more efficient in removing the adsorbate from solution. This is due to the greater availability of sorption sites at higher doses of the adsorbent. So, to estimate the optimum adsorbent dose for a particular system, the sample solution with a fixed adsorbate concentration should be made to come into contact with varying adsorbent doses till equilibrium is reached.

8.4.5.2 Contact Time

Adequate contact time between the adsorbent and the solution containing the adsorbate is essential for the adsorbent to approach equilibrium with the adsorbate. Each adsorbent particle has to purify a certain volume of liquid, so that the higher adsorbent dosages with less volume to treat per unit weight may reach equilibrium somewhat faster than the low dosages. To better estimate the optimum contact time for a particular system, a single adsorbent dosage should be made to come into contact with the solution for a certain period. Measurement of the concentration change over time in this system will show the effect of contact time.

8.4.5.3 Agitation Rate

There are essentially three consecutive mass transport steps associated with the adsorption of solutes from the solution by porous adsorbents, namely, bulk diffusion, film diffusion, and pore diffusion. The rate of adsorption is controlled by either film diffusion or pore diffusion, depending on the amount of agitation in the system. If relatively little agitation occurs between the fluid and the adsorbent, the surface film of the liquid around the particle will be thick and film diffusion will likely be the rate-limiting step. If adequate mixing is provided, the rate of film diffusion will increase to the point that pore diffusion may become the rate-limiting step. Pore diffusion is generally rate limiting for batch systems, as it provides a high degree of agitation.[34]

8.4.5.4 Effect of pH and Coexisting Ions

Generally, the important factors influencing the adsorption process and capacity of the adsorbent include pH, interference effects from other counter-ions in water, initial adsorbate concentration, and temperature. It was observed that many of the adsorbents developed for defluoridation have shown a reduction in their capacity while dealing with natural groundwater[38,39] compared with their laboratory performance in synthetic water; however, reasons for the same were not fully elucidated. Since nearly all natural waters contain traces of many chemical elements, these competing adsorbate compounds influence the solution nature of the sorption system. Only very few adsorbents exhibit controllable selectivity for specific compounds and so all the adsorbable compounds may compete for adsorption sites. This effect turns significant, as it can have dramatic consequences on the capacity of the adsorbent. So, in this study, the reasons for this behavior were also investigated. The pH of the system derives significance, as it controls the electrostatic interactions within the system, thereby affecting the adsorption capacity and removal rate. Experimental evidence suggests that the removal of fluoride by activated alumina (the popular adsorbent used for defluoridation) is highly sensitive to pH. In a study, the maximum adsorption capacity of the adsorbent, alum-impregnated activated alumina, was reported to be reduced by almost 10 times when the pH was increased from 4.0 to 9.0.[40] The reported maximum adsorption capacities and removal rates were usually in the acidic pH range of 5–6. However, the reported pHs of fluoride-rich groundwaters were generally in the alkaline range.[41] So, it can be rationally expected that those adsorbents having optimum pH in the acidic range will not be at their best in treating natural fluoride-rich waters. Thus, the effect of pH turns extremely significant in fluoride removal.

> Although it may be easy to adjust the pH for maximum removal in laboratory studies and waterworks, it is necessary to depend on the actual pH of raw water in domestic and small community treatments. So, for design and practical applications, it becomes necessary[2] for sorption profiles of the adsorbent to be established under prevailing field conditions. Since a reduction in adsorption capacity was generally expected (as demonstrated in the limited studies cited earlier) in treating natural groundwater, this study attempts to elucidate the factors responsible for the same.[2]

8.4.5.5 Temperature

The temperature of the adsorption process will affect both the rate and the extent of adsorption. The temperature of a solution has two major effects on adsorption. First, the rate of adsorption is usually increased at higher temperatures. This is due primarily to the increased rate of diffusion of adsorbate molecules through the solution to the adsorbent. Further, since solubility and adsorption are inversely related, the effect of temperature on solubility will naturally affect the extent of adsorption (or capacity of the adsorbent) onto a particular adsorbate. Hence, the temperature effects should be studied carefully and evaluated for possible effects.

The temperature dependence of equilibrium capacity for adsorption can be defined by the thermodynamic parameters enthalpy (ΔH°), entropy (ΔS°), and Gibbs free energy (ΔG°). These parameters are useful tools for delineating the nature of adsorption mechanisms. The change in the heat content of a system in which adsorption occurs, the total amount of heat evolved in the adsorption of a defined quantity of adsorbate on an adsorbent, is termed the *heat of adsorption* (ΔH°).[29] Standard Gibbs free energy (ΔG°), standard enthalpy (ΔH°), and entropy (ΔS°) changes for the adsorption process can be calculated from Equations 8.25 and 8.26:

$$\Delta G^0 = -RT \ln K \tag{8.25}$$

$$\ln K = \frac{\Delta S^0}{R} - \left(\frac{\Delta H^0}{R}\right)\left(\frac{1}{T}\right) \tag{8.26}$$

where T is the temperature (K), and R is the universal gas constant. When any spontaneous process occurs, there is a decrease in the Gibbs free energy. For significant adsorption to occur, the free energy changes of adsorption,

$\Delta G°$, must be negative. Further, there must also be a decrease in entropy, because the molecules lose at least one degree of freedom when adsorbed.[42]

8.4.5.6 Ionic Strength

Apart from pH, another important parameter in adsorption is the ionic strength. It is often stated in literature that an increase in ionic strength suppresses sorbate uptake as a result of the screening of electrostatic charge.[43] This fact is relevant to the way that the metal is electrostatically or covalently bound. Alternatively, other sorbable ions can compete with the cations of interest for binding with the adsorbent, thereby affecting adsorption. When direct evidence from microscopic data is absent, studying the influence of ionic strength is a simple approach to distinguish between the inner-sphere and outer-sphere surface complexes. If the adsorption is not affected by the variations of the ionic strength, an inner-sphere surface complexation should form; whereas if the adsorption is reduced with an increase in ionic strength (i.e., due to the competitive adsorption with counter-anions), an outer-sphere surface complexation is more likely.[44] In the absence of microscopic techniques (such as extended x-ray absorption fine structure [EXAFS]), investigation into the effect of ionic strength has been carried out by macroscopic studies.

8.4.6 Behavior of Adsorption Columns

When the solution containing the solute (i.e., fluoride) is introduced at the bottom of a packed bed containing the adsorbent media (ALC), most of the solute removal initially occurs in a narrow band at the bottom of the column, which is referred to as the *adsorption zone.*[10] As the column operation continues, the lower layers of the adsorbent bed become saturated with the solute and the adsorption zone progresses upward through the bed. Eventually, as the adsorption zone reaches the top of the column, the solute concentration in the effluent begins to increase. The loading behavior of the solute to be removed from the solution in a fixed bed containing the adsorbent media is shown by breakthrough curves and is usually expressed in terms of a normalized concentration that is defined as the ratio of effluent solute concentration to inlet solute concentration (C_t/C_0) as a function of time (t) or bed volumes for a given bed height (Z).

The breakthrough curve, developed from the column studies, basically portrays the dynamic sorptive responses of the adsorbent. The point at which the effluent fluoride concentration reaches 1 mg/L is taken as *breakthrough point* and that corresponding to 90% of influent concentration (C_t/C_0 = 0.90) is considered *point of exhaust*. The capacity of the column up to the point of breakthrough will represent the *breakthrough capacity*, which is indicative of the column adsorption capacity in a single-column operation. In series column operations, the effluent from the last column represents the effluent of the desired quality. So, all columns except the last one can be run up to the exhaustion point, as these columns still contain unused adsorbents. In such cases, *breakthrough capacity* serves as the minimum capacity of the adsorbent in the column ($q_{\mathrm{min,col}}$), and the capacity up to the exhaust will be its total or maximum capacity (q_{col}). The value of $q_{\mathrm{min,col}}$ turns useful in the design of domestic defluoridation units (DDUs), which usually involve a single-chamber use; however, for field-scale community applications, the involvement of a number of columns q_{col} will be appropriate.

The total quantity of fluoride adsorbed (F_{tot}) in the column for a given feed concentration (C_0) and flow rate (Q, l/h) can be found by calculating the area above the breakthrough curve by integrating the adsorbed fluoride concentration ($C_{\mathrm{ad}} = C_0 - C_t$) versus time t (h) plot as follows[2]:

$$F_{\mathrm{tot}} = Q \int_{t=0}^{t=et} C_{\mathrm{ad}}\, dt \tag{8.27}$$

Similarly, the quantity of fluoride adsorbed up to the breakthrough (F_{b}) is as follows:

$$F_{\mathrm{b}} = Q \int_{t=0}^{t=bt} C_{\mathrm{ad}}\, dt \tag{8.28}$$

The minimum and maximum adsorption capacity of ALC in the sorptive filtration system can be calculated as follows:

$$q_{\mathrm{min,col}} = \frac{F_{\mathrm{b}}}{M} \tag{8.29}$$

$$q_{\mathrm{col}} = \frac{F_{\mathrm{tot}}}{M} \tag{8.30}$$

where M is the mass of the adsorbent in the column.

The prediction of column breakthrough or the shape of the adsorption wave front, which determines the operation life span of a bed and regeneration times, is the most important criterion in the successful design of fixed bed adsorption systems. However, it is innately difficult to develop a model that accurately describes the dynamic behavior of adsorption in a fixed bed system. The process does not operate in a steady state, as the concentration of the adsorbate changes as the feed moves through the bed. The fundamental transport equations for a fixed bed are those of material balance between the solid and the fluid. The equation of material mass balance can be stated as follows: input flow = output flow + flow inside pore + matter adsorbed onto the bed.[45]

The material mass balance equation for this system can be expressed mathematically as follows:

$$QCo = QC_t + V_{\mathrm{p}}\frac{dC}{dt} + m\frac{dq}{dt} \tag{8.31}$$

where (QC_0) is the inlet flow of solute to the column (mg/min); (QC_t) is the outlet flow of solute leaving the column (mg/min); V_{p} is the porous volume (l) ($V_{\mathrm{p}} = V/(1-\varepsilon)$, where V is the bulk volume (l) and ε is the void fraction in the bed; $V_p(dC/dt)$ is the flow rate through the column bed depth (mg/min); and $m(dq/dt)$ is the amount of solute adsorbed onto the sorbent media (mg/min), where m is the mass of the adsorbent (g) and (dq/dt) is the adsorption rate (mg/g/min).[45]

From the relationship just cited (Equation 8.31), it is evident that the linear flow rate ($u = Q/A$, where A is the column section area), the initial solute concentration, and the adsorption potential are the determining factors of the balance for a given column bed depth. Therefore, it is necessary to examine these parameters and to estimate their influence in order to optimize the fixed bed column adsorption process. However, these equations that are derived to model the fixed bed adsorption system with theoretical vigor are differential in nature and usually require complex numerical methods to be solved.[45] Because of this, various simple numerical models have been developed to predict the dynamic behavior of the columns and some of these models have been discussed in this study. The prediction and analysis of the dynamic behavior of the column was carried out with Hutchin's bed depth service time (BDST) model, Thomas model, Yoon–Nelson model, Clark model, Wolborska model, and Bohart and Adams model.

8.4.7 Analysis and Modeling of Breakthrough Profile

For a given bed depth, the service times of a unit plant are correlated with the initial sorbate concentration, flow rate, and adsorption capacity of the adsorbent to be used. So, obtaining a meaningful and reliable loading capacity of the adsorbent turns crucial in efficient process design and operation. This necessitates a careful evaluation and analysis of the experimental data, to predict the effect of variations in operational parameters of the sorption process, through modeling.

8.4.7.1 Hutchins BDST Model

The BDST model proposed by Hutchins in 1973 is based on the assumptions that intra particular diffusion and external mass resistance are negligible and that adsorption kinetics is controlled by the surface chemical reaction between the solute in the solution and the unused adsorbent.[46] A linear relationship between the column depth (Z) and service time (t) was proposed as follows:

$$t = \frac{N_0}{C_0 u} Z - \frac{1}{KC_0} \ln\left(\frac{C_0}{C_t} - 1\right) \tag{8.32}$$

where C_0 is the initial solute concentration (mg/L), C_t is the desired solute concentration at breakthrough (mg/L), K is the adsorption rate constant (L/mg/min), N_0 is the adsorption capacity (mg/L), Z is the depth of the sorbent bed (cm), u is the linear flow velocity of feed to bed (cm/min), and t is the service time of column under the conditions mentioned (min). The dynamic bed capacity (N_0) and the adsorption rate constant (K) can be evaluated by the linear regression of the following straight-line relationship:

$$t = aZ + b \tag{8.33}$$

where slope

$$a = \frac{N_0}{C_0 u} \tag{8.34}$$

intercept

$$b = -\frac{1}{KC_0}\ln\left[\frac{C_0}{C_t} - 1\right] \tag{8.35}$$

The BDST model is a useful tool for comparing the performance of columns operating under different process variables. If there is a change in the initial solute concentration C_0 to a new value C_0', the new values of a' and b' can be, respectively, obtained from the slope and the intercept according to the relations proposed by Hutchins:

$$a' = a\left(\frac{C_0}{C_0'}\right) \tag{8.36}$$

$$b' = b\left(\frac{C_0}{C_0'}\right)\frac{\ln\left(C_0'/C_b - 1\right)}{\ln\left(C_0/C_b - 1\right)} \tag{8.37}$$

When the linear flow rate is changed from u to u', the new gradient a' can be calculated as follows:

$$a' = a\left(\frac{u}{u'}\right) \tag{8.38}$$

The intercept remains unchanged, because it depends on only the inlet solute concentration C_0. This is useful to scale up the process for other flow rates without further experimental run. Also, at 50% breakthrough ($C_t/C_0 = 0.5$), the term b in Equation 8.35 becomes zero, and Equation 8.33 is reduced to the following:

$$t_{50} = aZ \tag{8.39}$$

So, if the sorption process follows the BDST model, the plot of t against Z at 50% breakthrough will represent a straight line passing through the origin.

8.4.7.2 Thomas Model

The Thomas model[47] is another most general and widely used model in column performance theory. This model is derived by assuming Langmuir kinetics of adsorption–desorption with no axial dispersion and that the rate-driving force obeys second-order reversible reaction kinetics. The data obtained in fixed bed column studies are used to calculate the maximum solid-phase concentration of sorbate on the sorbent and the adsorption rate constant. The expression by Thomas for an adsorption column is given as follows:

$$\frac{C_t}{C_0} = \frac{1}{1 + \exp\left(\frac{k_{\mathrm{Th}} q_{\mathrm{Th}} M}{Q} - k_{\mathrm{Th}} C_0 t\right)} \quad (8.40)$$

where k_{Th} is the Thomas rate constant (mL/min/mg), q_{Th} is the equilibrium sorbent uptake per gram of the adsorbent (mg/g), M is the amount of adsorbent in the column (g), C_0 is the influent sorbate concentration (mg/L), C_t is the effluent concentration at time t (mg/L), Q is the flow rate (mL/min), and t is the sampling time. The value of C_t/C_0 is the ratio of effluent and influent sorbate concentrations.

The linearized form of the Thomas model is as follows:

$$\ln\left(\frac{C_0}{C_t} - 1\right) = \frac{k_{\mathrm{Th}} q_{\mathrm{Th}} M}{Q} - k_{\mathrm{Th}} C_0 t \quad (8.41)$$

The kinetic coefficient k_{Th} and the adsorption capacity of column q_{Th} can be determined from a plot of $\ln[(C_0/C_t) - 1]$ against t at a given flow rate.

8.4.7.3 Yoon–Nelson Model

The Yoon–Nelson model[48] is based on the assumption that the rate of decrease in the probability of adsorption for each adsorbate molecule is proportional to the probability of adsorbate adsorption and the probability of adsorbate breakthrough on the adsorbent. This model requires no detailed data regarding the characteristics of adsorbate, the type of adsorbent, and physical properties of the adsorption bed and so it is less complicated. The model is expressed as follows:

$$\frac{C_t}{C_0} = \frac{1}{1 + \exp\left[k_{\mathrm{YN}}(\tau - t)\right]} \quad (8.42)$$

where k_{YN} is the rate constant (min^{-1}), τ is the time required for 50% adsorbate breakthrough (min), and t is the breakthrough (sampling) time (min). The linearized form of the Yoon–Nelson model is as follows:

$$\ln\left(\frac{C_t}{C_0 - C_t}\right) = k_{\mathrm{YN}}\, t - \tau k_{\mathrm{YN}} \quad (8.43)$$

The parameters k_{YN} and τ may be determined from the plot of $\ln[C_t/(C_0 - C_t)]$ versus sampling time (t).

8.4.7.4 Clark Model

The model developed by Clark (1987)[49] was based on the use of a mass-transfer concept in combination with the Freundlich isotherm:

$$\left(\frac{C_0}{C_t}\right)^{n-1} - 1 = Ae^{-rt} \tag{8.44}$$

where n is the Freundlich parameter, and A and r are the Clark constants:

$$A = \exp\left(\frac{K_c N_0 Z}{u}\right) \text{and } r = K_c C_0 \tag{8.45}$$

Linearizing Equation 8.44:

$$\ln\left(\left[\frac{C_0}{C_t}\right]^{n-1} - 1\right) = \ln A - rt \tag{8.46}$$

From a plot of $\ln[(C_0/C_t)^{n-1} - 1]$ versus time, the values of r and A can be determined from its slope and intercept, respectively.

8.4.7.5 Wolborska Model

The Wolborska model[50] is used for describing adsorption dynamics by using mass transfer equations for diffusion mechanisms in the low-concentration ranges of the breakthrough curve. The mass transfer in the fixed bed sorption is described by the following equations:

$$\frac{\partial C_b}{\partial t} + u\left(\frac{\partial C_b}{\partial Z}\right) + \left(\frac{\partial q}{\partial t}\right) = D\left(\frac{\partial^2 C_b}{\partial^2 Z}\right) \tag{8.47}$$

$$\frac{\partial q}{\partial t} = -v\left(\frac{\partial q}{\partial Z}\right) = \beta(C_b - C_s) \tag{8.48}$$

where C_b is the fluoride concentration in solution (mg/L), C_s is the fluoride concentration at the solid/liquid interface (mg/L), β is the kinetic coefficient of the external mass transfer (h^{-1}), v is the migration rate (cm/min), D is the axial diffusion coefficient (cm^2/min), and q is the fluoride concentration on the sorbent at any time t (mg/L). With some assumptions previously described by Wolborska: $C_s \ll C_b$, $v \ll u$, and axial diffusion is negligible ($D \rightarrow 0$ as $t \rightarrow 0$), the solution can be approximated to[50]

$$\ln\left(\frac{C_t}{C_0}\right) = \frac{\beta C_0 t}{N_0} - \frac{\beta Z}{u} \tag{8.49}$$

8.4.7.6 Bohart and Adams Model

Bohart and Adams (1920)[51] proposed the fundamental equations describing the relationship between C_t/C_0 and t for the quantitative description of a continuous flow fixed bed system. It was assumed that adsorption rate is proportional to both the residual capacity of the adsorbent and the concentration of the sorbate. The Bohart and Adams model is generally used only for describing the initial part of the breakthrough curve ($C_t/C_0 \sim 0.5$). The mass transfer rates obey the following equations[52]:

$$\frac{\partial q}{\partial t} = -k_{AB} q\, C_b \tag{8.50}$$

$$\frac{\partial C_b}{\partial Z} = -\frac{k_{AB}}{u} q C_b \tag{8.51}$$

where k_{AB} is the Adams and Bohart kinetic constant (L/mg/h), q is the fluoride concentration in the sorbent at any time t (mg/L), C_b is the fluoride concentration in solution (mg/L), and u is the linear flow velocity of feed to bed (cm/min). The solution of the differential equations cited earlier with the following assumptions of a low concentration field ($C_t < 0.15C_0$) and that when $t \to \infty$, $q \to N_0$ (its saturation concentration) renders a linear relationship between its parameters as follows:

$$\ln\left(\frac{C_t}{C_0}\right) = k_{AB} C_0 t - k_{AB} N_0 \frac{Z}{u} \tag{8.52}$$

From this equation, values describing the characteristic operational parameters of the column can be determined from a plot of $\ln(C_t/C_0)$ against t, at a given bed height and flow rate. In all cases, the average percentage errors (APE) between the experimental and predicted values were calculated as follows[1]:

$$\text{APE\%} = \frac{\sum_{i=1}^{N} \left| \frac{C_{t(\exp)} - C_{t(\text{theo})}}{C_{t(\exp)}} \right|}{N} \times 100 \tag{8.53}$$

where N is the total number of samples.

8.4.8 Regeneration

The regeneration capacity of an adsorbent also plays an important role in making sorptive systems economical. The exception is where there is very long adsorption or loading cycles due to very low concentrations of solute in the inlet feed; this type of system usually uses the adsorbent only once on a *throw away* basis and safe disposal becomes a problem. If very large quantities of adsorbent are involved, regeneration and reuse becomes necessary. Regeneration of adsorbents can be accomplished with heat, chemical change, or solvent action. Each of these methods has advantages, and disadvantages when applied to specific adsorbates, adsorbents, and systems. The regeneration methods using solvents are relatively straightforward and are commonly used to determine the working capacity of an adsorbent. Thermal treatment is more difficult to perform and is related to full-scale performance. Very few regeneration methods can be economically operated to 100% efficiency. Generally, there will be a reduction in the capacity of the adsorbent due to successive regeneration.

Since the solute is bound to the adsorbent by physical and/or chemical forces, the regeneration procedure must develop conditions that these forces should overcome. This can be accomplished either by subjecting the system to conditions where the attractive forces for the adsorbed solute by the regenerating medium are greater than the adsorbent attractive forces or by chemically changing the solute so that the binding forces are neutralized. An elution curve, which plots the concentration of the sorbate in the regenerant as a function of time or regenerant volume, will describe the efficiency of regeneration. The shape of the regeneration curve can be influenced by regenerant concentration, temperature, and flow rate. At the end of the regeneration, the void spaces in the adsorbent bed are filled with the regenerant solvent. Before returning the adsorbent bed back into service, the regenerant must be adequately rinsed from the adsorbent grains.

8.5 Results and Discussion

8.5.1 Characterization of the Adsorbent

The physical and chemical properties of an adsorbent play an important role in determining its performance in terms of its sorption capacity. The adsorbent used in this study, hardened granules of high alumina cement, is a composite mixture of Al–Ca–Si–Fe containing substances. The adsorbent is abbreviated as ALC. It is a variety of commercially available cement with a high content of alumina (Al_2O_3). XRD analysis (Figure 8.2) identified the presence of compounds such as

> aluminum oxide (Al_2O_3), aluminum iron silicon ($Al_{0.7}Fe_3\ Si_{0.3}$), magnetite (Fe_3O_4), calcium iron oxide ($Ca_4Fe_9O_{17}$), calcium aluminum oxide ($Ca_3Al_{10}O_{18}$), silicon oxide (SiO_2), calcium aluminum silicate ($Ca_{0.88}Al_{1.77}Si_{2.23}O_8$), and wollastonite ($CaSiO_3$) along with many other trace compounds.[3]

The physical parameters and major chemical constituents of ALC media identified by EDX are shown in Table 8.1. The surface texture of ALC was observed by SEM studies. Figure 8.3 shows the SEM photograph of ALC at a magnification of 500 μm. Figure 8.4 (at 10 μm magnification) shows a rough and a highly porous surface texture that may be effective for high sorptive removal.

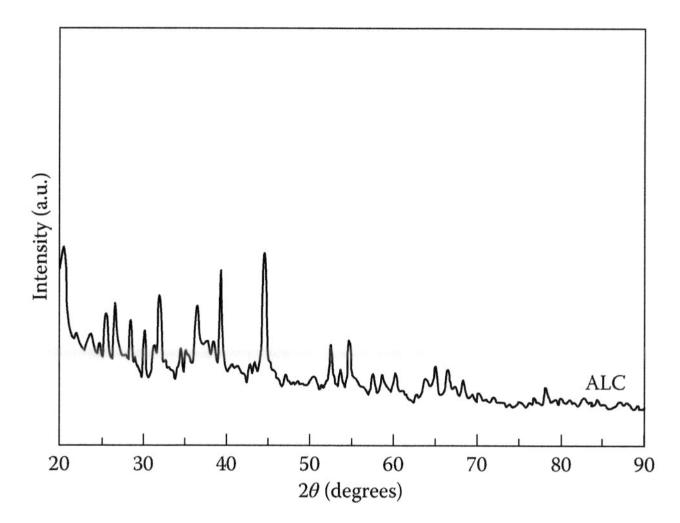

FIGURE 8.2
The x-ray diffractogram of ALC. (Modified from Ayoob, S., Gupta, A.K., Bhakat, P.B. and Bhat, V.T., *Chem. Eng. J.*, 140, 6–14, 2008.)

TABLE 8.1
Properties of ALC Media

Properties	Quantitative Value
Geometric mean size (mm)	0.212
Bulk density (g/cm^3)	2.33
Al_2O_3 (%)	78.49
CaO (%)	15.82
SiO_2 (%)	5.39
Fe_2O_3 (%)	0.30
pH of the PZC	11.32
BET surface area (m^2/g)	4.385

Source: Ayoob, S. and Gupta, A.K., *Chem. Eng. J.*, 150, 485–491, 2009.

8.5.2 Kinetics Studies

The adsorption rate is strongly influenced by several parameters related to the state of the solid, generally having a very heterogeneous reactive surface, and to the physicochemical conditions under which adsorption is carried out.

8.5.2.1 Agitation Rate

The influence of agitation speed on the sorption of fluoride onto ALC was studied by changing the speed of agitation from 120 to 280 rpm. Figure 8.5 shows that the sorption was influenced by the rate of agitation. The removal increases from 64.99% to 78.49% and from 74.04% to 90.09% with

FIGURE 8.3
Scanning electron microscopic (SEM) photograph of ALC particles at 500 μm magnification.

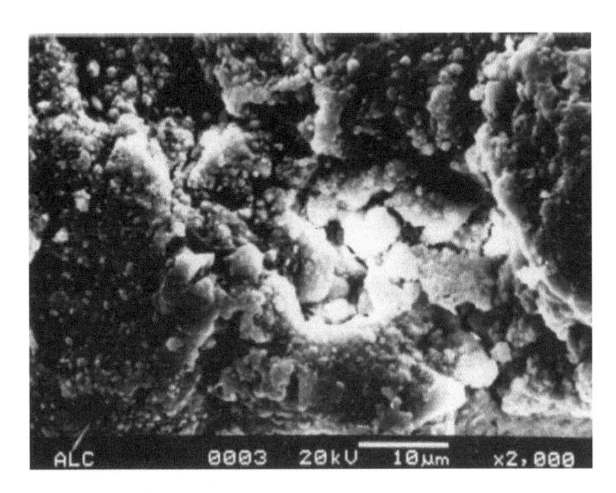

FIGURE 8.4
SEM photograph of ALC particles at 10 μm magnification.

an increase in agitation speed from 120 to 240 rpm in natural and synthetic waters, respectively. The increasing agitation rate decreases the boundary layer resistance to mass transfer in the bulk and increases the driving force of the fluoride ions. This may indicate that film diffusion does not dominantly control the overall adsorption process.[53] Further sorption studies were carried out at an agitation rate of 230 ± 10 rpm.

8.5.2.2 *Adsorbent Dosage*

The response of the adsorbent to different adsorbent dosages is as shown in Figure 8.6. The observed increase in removal efficiency with an increase in

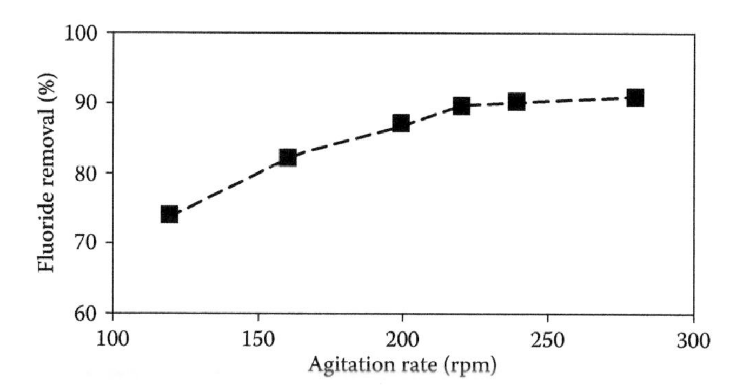

FIGURE 8.5
Effect agitation rates on fluoride sorption onto ALC.

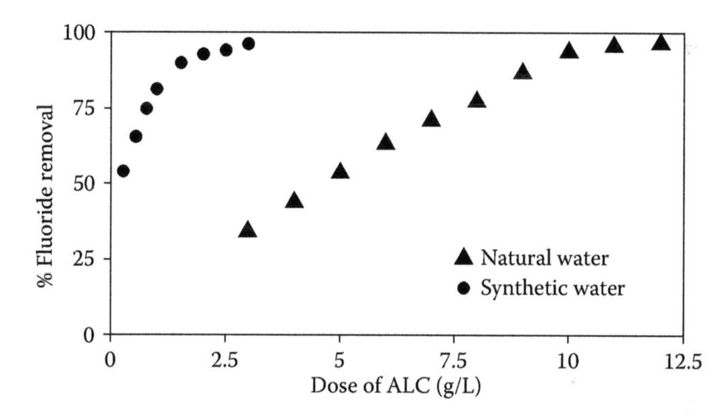

FIGURE 8.6
Effect of dose variations of ALC on fluoride removal percentage. (From Ayoob, S. and Gupta, A.K., *Chem. Eng. J.*, 150, 485–491, 2009.)

ALC doses implies that the process is dependent on the availability of sorptive binding sites. This sorption pattern indicates the predominance of surface adsorption, since both the internal and external sorption sites are found to increase with higher adsorbent dosage.[54]

> It is postulated that at low adsorbent dosages all types of sites are entirely exposed to adsorption, and the surface may become saturated faster. However, at higher adsorbent dosages, the availability of higher energy sites may be reduced and a larger fraction of lower energy sites may get occupied. This results in an overall decrease in binding energy of the surface, and a reversible type of process exists between the fluoride ions attached to low energy sites and those present in bulk solution. This could be the reason for the observed increase in uptake with an increase in adsorbent dosage up to a certain stage and almost a constant uptake thereafter.[2]

8.5.3 Kinetic Profile of Fluoride Uptake

The kinetic curve of fluoride removal sorption (Figures 8.7 and 8.8) displays a rapid uptake for the initial few minutes and is followed by a slow phase. The removal process reaches an equilibrium stage after around 150 min with negligible removal after 3 h. Within the first 10 minutes itself, around 70%–75% fluoride gets removed. The maximum uptake of ALC at equilibrium in groundwater (fluoride concentration of 8.65 mg/L) was found to be 0.806 mg/g.[2,3] The increase in adsorption capacity in the synthetic system at higher fluoride concentrations, signifying heterogeneous sorptive surfaces, may be ascribed to high intramolecular competitiveness to occupy the unsaturated lower energetic surface sites.[7] This biphasic fluoride sorption behavior suggests the key role of mass transfer in the removal process. The initial rapid uptake may indicate surface-bound sorption and precipitative removal. The slow second phase of fluoride sorption may be ascribed to the long-range diffusion of fluoride ions onto the interior pores of ALC.[55]

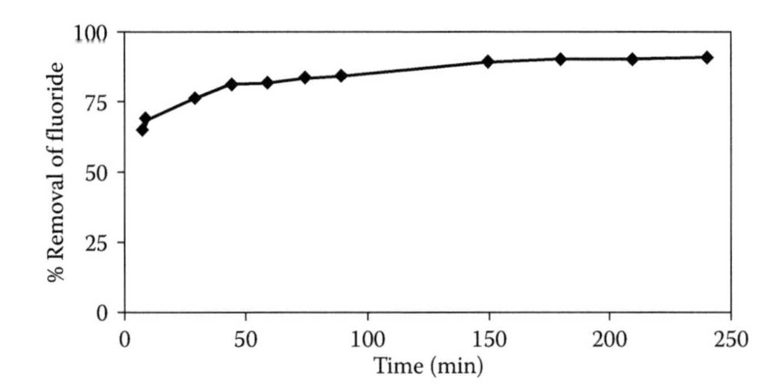

FIGURE 8.7
Kinetic curve of fluoride sorption onto ALC. (Modified from Ayoob, S. and Gupta, A.K., *Chem. Eng. J.*, 133, 273–281, 2007.)

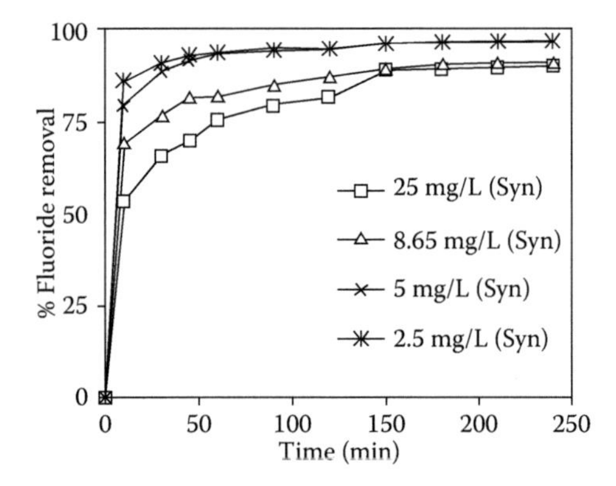

FIGURE 8.8
Kinetic curve of fluoride sorption onto ALC at different concentrations.

Kinetic models were applied to examine and describe the dynamics of sorption of fluoride onto ALC. Kinetic modeling was carried out by the pseudo-first-order model, pseudo-second-order model, intra particle surface diffusion model, and the Elovich model; a comparison of the best-fit sorption mechanisms was also made.

8.5.3.1 Pseudo-First-Order Model

The linearized form of the pseudo-first-order model indicates that a linear fit between $\ln(q_e - q_t)$ and contact time (t) demonstrates that the reaction may follow a pseudo-first-order reaction.[3] The plots of $\ln(q_e - q_t)$ versus t for synthetic water is shown in Figure 8.9. This reasonable linear fitting of the kinetic data in most portions of the contact time shows the possibilities of a diffusion-controlled sorption mechanism,[15] though the fitting in the initial portions is relatively poor. The correlation coefficients were found to be 0.975 and 0.959 in synthetic and natural waters, respectively. The pseudo-first-order rate constant k_{s1} for synthetic and natural water that was calculated from the slopes of the linear plots were found to be 0.0215 and 0.0164 min^{-1}, respectively.[3]

8.5.3.2 Pseudo-Second-Order Model

A linear fit between t/q_t and contact time (t) indicates that the reaction is of pseudo–second order. The plots of t/q_t versus t for synthetic water is shown in Figure 8.10. Linear regression curves show a good correlation, as R^2 is 0.999 in synthetic waters and 0.998 in natural waters. This shows that the pseudo-second-order model provides a better fit to the sorption kinetics data of fluoride onto ALC over the whole range of the sorption process.[3] The pseudo-second-order constants k and h were calculated from the intercept and slope of the lines obtained by plotting t/q_t against t. The initial sorption rates (h) of sorption were found to be 0.9706 and 0.1877 mg/g/h in synthetic and natural waters with corresponding values of rate constants (k) as 0.0625 and 0.2891 g/mg/min.

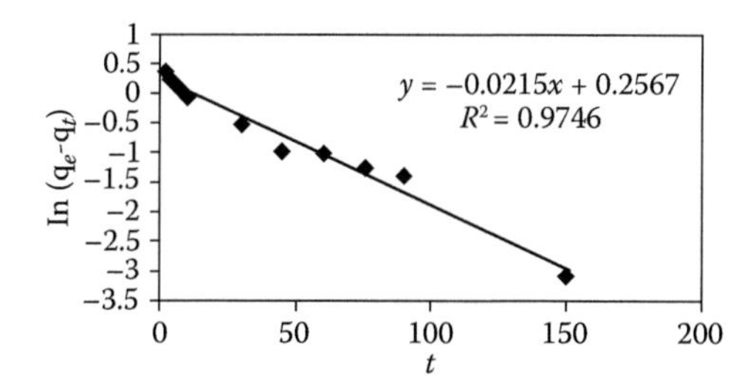

FIGURE 8.9
Pseudo-first-order kinetic fit of fluoride sorption onto ALC.

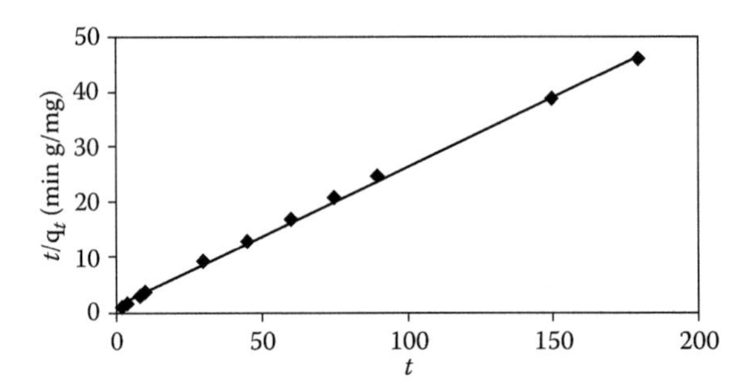

FIGURE 8.10
Pseudo-second-order kinetic fit of fluoride sorption onto ALC.

8.5.3.3 Intra Particle Surface Diffusion Model

The use of the intra particle surface diffusion model has been widely explored to elucidate the mechanism of the sorption process. It is suggested that if the plot of q versus $t^{1/2}$ renders a straight line, the sorption process is controlled by intra particle diffusion. If it does not pass through the origin, it indicates that the intra particle diffusion is not the only rate-limiting step. It further suggests that the process is *complex*, with more than one mechanism limiting the rate of sorption. In this study, though the plot of adsorbate uptake versus the square root of time (Figure 8.11) could be represented by such a linear relationship ($R^2 > 0.899$), it was found to not pass through the origin. This indicates that intra particle diffusion is involved in the sorption process. However, it is not the only rate-limiting mechanism and some other mechanisms may also get involved. It was pointed out that such a deviation of the straight line from the origin may be due to the difference in the rate of mass transfer in the initial stage of sorption.[56,57] The values of the intercept C provide information about the thickness of the boundary layer, that is, the resistance to the external

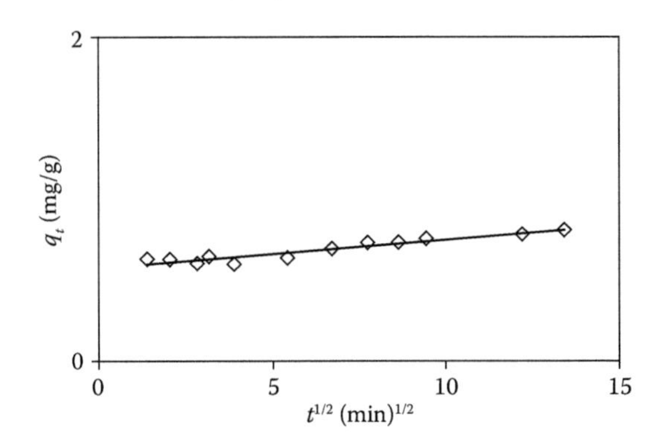

FIGURE 8.11
Intra particle diffusion model kinetic fit of fluoride sorption onto ALC.

mass transfer. The larger the intercept is, the higher the external resistance; namely, any increase in the intercept indicates the abundance of solute adsorbed on the boundary layer. The slope shown in Figure 8.11 was used to find the rate parameter (k_p) of the model. The application of this model to the natural and synthetic systems of this study rendered values of R^2, k_p, and C as 0.912, 0.1172 mg/g/h$^{1/2}$, and 2.5261 mg/g respectively, in synthetic water with corresponding values of 0.899, 0.0164 mg/g/h$^{1/2}$, and 0.5843 mg/g in natural water.

8.5.3.4 Elovich Model

The Elovich equation has been applied in aqueous systems to describe adsorption and desorption reactions. The Elovich equation can be derived from either a diffusion-controlled process or a reaction-controlled process.

> If the Elovich equation is based on adsorption on an energetically heterogeneous surface, the parameter β is related to the distribution of activation energies. However, in the diffusion control model, it is a function of particle and diffusion coefficients.[19]

As per the model, if it follows the Elovich equation, the kinetic results will be linear on a q_t versus ln t plot. The kinetic curve of sorption demonstrates excellent fitting (Figure 8.12) in synthetic systems (R^2 = 0.996) and natural systems (R^2 = 0.838). The fitting to the Elovich kinetics pattern indicates that the rate-determining step is diffusion in nature.[58]

On comparing the fitting of the applied kinetic models, it is evident that the kinetic profile could be best modeled by pseudo–second order, which is indicative of a chemisorptive rate-limiting step. The model could also excellently predict the equilibrium adsorption capacity of ALC in natural and synthetic water as 0.806 and 3.941 mg/g compared with corresponding experimental values of 0.803 and 3.912 mg/g.

8.5.3.5 Arrhenius Equation

The value of energy of activation (E_a) obtained from Arrhenius equation could also be a useful kinetic parameter in assessing rate-limiting steps. Low E_a values indicate diffusion-controlled transport and physical adsorption processes, whereas higher E_a values indicate chemical reactions or surface-controlled processes. Some of the assigned values of E_a (kJ/mol) include 8–25 to physical adsorption, less than 21 to aqueous diffusion, 20–40 to pore diffusion, and greater than 84 to ion exchange.[23] The plot of ln K versus $1/T$, which is required to calculate the activation energies, is shown in Figure 8.13.

The activation energies of the sorption process were calculated by multiplying the slope of the curves by the value of R. Accordingly, the values of E_a were found to be 17.67 and 20.12 kJ/mol in synthetic and natural waters,

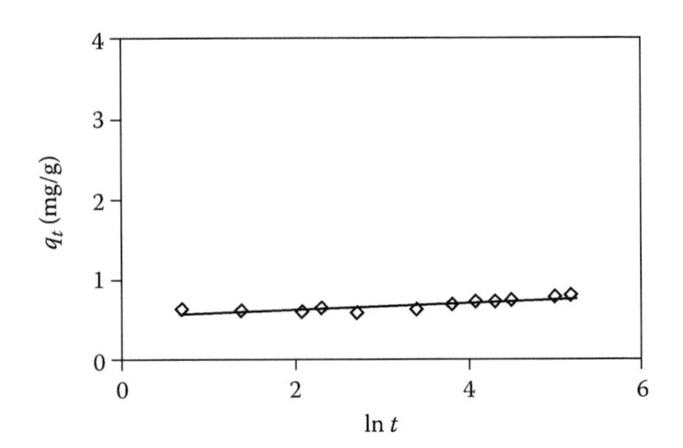

FIGURE 8.12
Elovich model kinetic fit of fluoride sorption onto ALC.

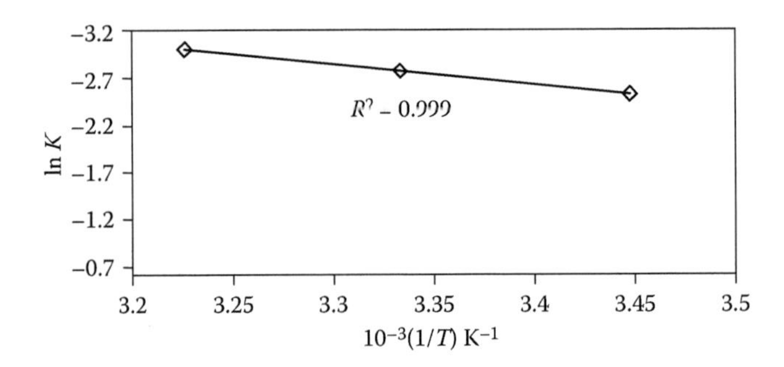

FIGURE 8.13
The plot of ln K versus $1/T$ of fluoride sorption onto ALC in synthetic water. (Modified from Ayoob, S. and Gupta, A.K., *Chem. Eng. J.*, 150, 485–491, 2009.)

respectively.[3] This value of activation energies signifies the role of diffusion-controlled transport and physical adsorption processes in the rate limiting of fluoride sorption onto ALC.[3]

8.5.4 Elucidation of Rate-Limiting Step

In this study, the following procedure was adopted to elucidate the rate-limiting step[19]:

> (1) The adsorption patterns should be sensitive to inert electrolyte concentrations or pH if the inter particle diffusion is rate limiting. (2) The desorption pattern will follow the same two-stage pattern as that of sorption if intra particle diffusion is rate limiting. (3) If both inter particle and intra particle diffusion are not rate controlling, the surface reactions will be rate limiting.[3]

So, to ascertain whether inter particle diffusion is rate limiting, the effects of pH and electrolyte concentrations were investigated. ALC exhibited consistent removal (Figure 8.14) in the pH range of 3–11.5, and it was reduced slightly thereafter by around 10% at pH 12.[3]

> The pH dependence of fluoride sorption onto ALC could be well explained in terms of its pH_{ZPC} (11.32). The pH_{ZPC} indicates where the net surface charge on the adsorbent is zero. When pH < pH_{ZPC}, the net surface charge on the solid surface of ALC is positive due to adsorption of excess H^+, which favors adsorption due to coulombic attraction. At pH > pH_{ZPC}, the net surface charge is negative due to desorption of H^+ and adsorption must compete with coulombic repulsion. The consistent fluoride removal in the range of 3–11.5 could be due to the combined effect of both chemical and electrostatic interactions between the oxide surfaces and fluoride ion. The observed reduction in fluoride adsorption above pH 11.5 may suggest that the strong negative surface charge developed at this pH may cause repulsion for the available adsorption sites. However, the considerable potential of the adsorbent above its pH_{ZPC} of 11.32 may be attributed to the predominance of specific adsorption due to chemical interactions.[3]

As shown in Figure 8.14,

> fluoride sorption is also found to be unaffected by electrolyte concentrations in the range of 10^{-4} M $NaNO_3$ to 10^{-1} M $NaNO_3$, though the marginal increase observed at higher concentrations may be due to compression of the electrostatic double layer.[3,59] Since the sorption process is unaffected by inert electrolyte concentrations and pH over a wide range, it can be suggested that the fluoride removal percentage, and hence the surface coverage, approaches almost the same value within the range of inert electrolyte concentrations tested (10^{-4} to 10^{-1} M), and the rate of sorption is unaffected by the effective particle size or diffusive path lengths.[3]

This response to pH and inert electrolyte concentrations

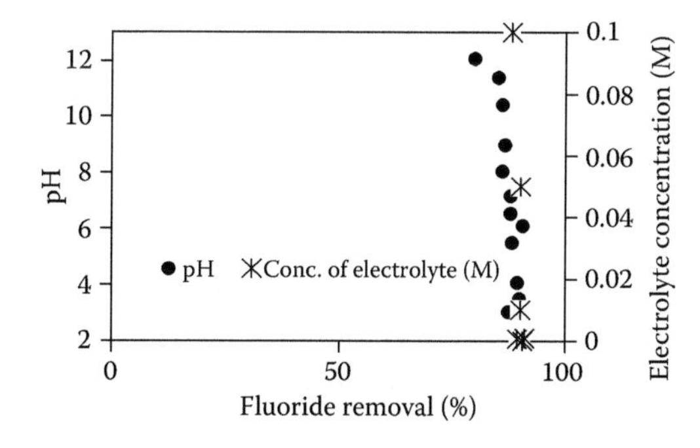

FIGURE 8.14
Effect of pH and inert electrolyte concentrations on the fluoride sorption onto ALC. (Modified from Ayoob, S., Gupta, A.K., Bhakat, P.B. and Bhat, V.T., *Chem. Eng. J.*, 140, 6–14, 2008.)

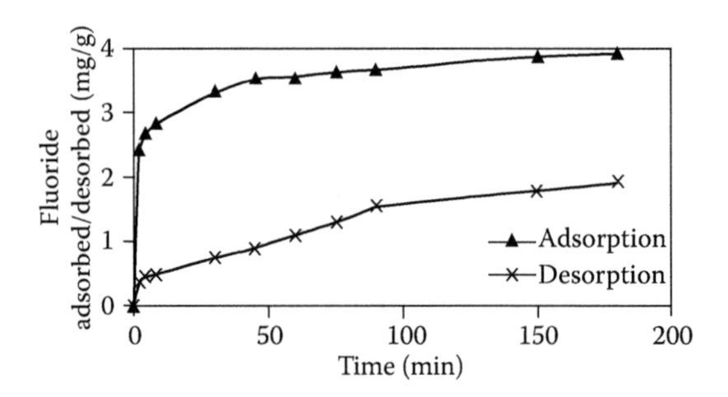

FIGURE 8.15
The sorption and desorption patterns of fluoride by ALC. (Modified from Ayoob, S., Gupta, A.K., Bhakat, P.B. and Bhat, V.T., *Chem. Eng. J.*, 140, 6–14, 2008.)

> clearly demonstrates that inter particle or external diffusion is not the rate-determining step. Since the possibility of inter particle or external diffusion is ruled out, the other possibility is that of intra particle diffusion. Since the presence and shape of the pores in an adsorbent does not change with pH or inert electrolyte concentrations, and only depends on its crystal properties, the desorption behavior of the adsorbent gives enough indication as to whether intra particle diffusion is rate limiting.[3]

However, in batch desorption studies, it was observed that concentrations of fluoride increased slowly with time with nearly 55% and 63% of the fluoride not being desorbed in synthetic and natural systems, respectively. The sorption and desorption patterns are shown in Figure 8.15.

> The desorption pattern in column studies also showed a poor response with much of the fluoride not being desorbed using 10% NaOH, even after 20 bed volumes. So, it is clear that the sorption and desorption patterns are not identical, demonstrating that the diffusion into ALC (i.e., intra particle diffusion) is not the rate-limiting process. Since neither inter particle nor intra particle diffusion is rate limiting, it indicates that the slow sorption phase in fluoride sorption is not the result of diffusion. So by elimination, it could be plausible that it is due either to the heterogeneity of the surface site-binding energy or to other reactions controlling fluoride removal from solution.[3]

8.5.5 Fluoride Removal Mechanism

The mechanisms of sorptive removal of fluoride may include specific or non specific adsorption or both. Experimental evidence suggests that anions that form inner-sphere complexes coordinate directly with the oxide surface without getting influenced by the ionic strength.[60] Since the sorption of fluoride onto ALC is relatively independent of the electrolyte concentrations and pH, the mechanism of removal can be ascribed to inner-sphere complex formations.

> The poor desorption characteristics of ALC suggest stronger adsorption of fluoride ions, further supporting the formations of inner-sphere complexes. As cited earlier, the SEM analysis observed that Al_2O_3 was the most prominent metal oxide in ALC. Further, the XRD analysis (Figure 8.16), which was carried out to identify the morphological structure and the extent of crystallinity of the adsorbent, shows multiple peaks, thus indicating the presence of various oxides and heterogeneous surface sites for sorption.[3]

On hydration, the metal ions on the oxide surface complete their coordination shells with OH groups. Depending on the pH, these OH groups can bind or release H^+, resulting in the development of a surface charge as follows:

$$MOH + H^+ = MOH_2^+ \tag{8.54}$$

$$MOH = MO^- + H^+ \tag{8.55}$$

where M is the metal (Al, Si, Fe, etc.) and MOH_2^+, MOH, and MO^- are positive, neutral, and negative surface hydroxo and oxo groups, respectively.[3]

The adsorption properties of the metal oxides are due to the presence of these OH_2^+, OH, and O^- surface functional groups,[61] as they dictate the number of reactive sites. Accordingly, a surface complex formation model is proposed (ligand-exchange model) to describe fluoride adsorption on metal oxides as follows[3,36]:

$$MOH + F^- = MF + OH^- \quad (pH > 7) \tag{8.56}$$

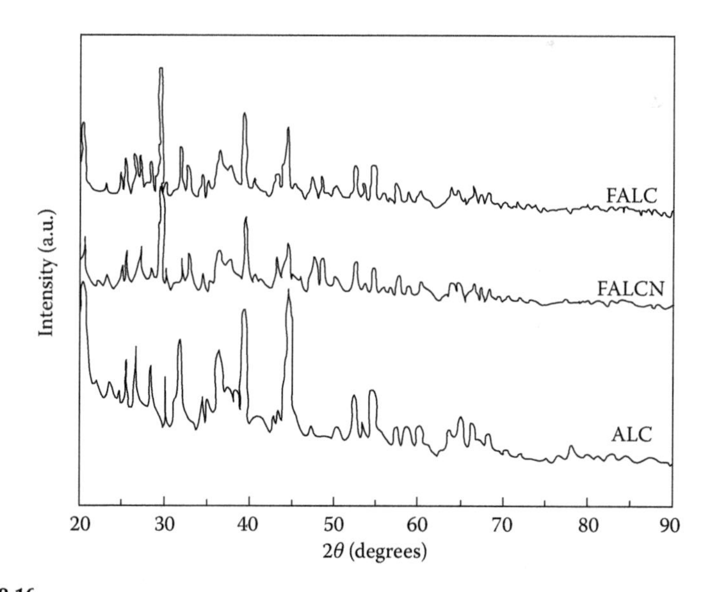

FIGURE 8.16
XRD pattern of the adsorbent before sorption (ALC) and after sorption of fluoride in natural (FALCN) and synthetic (FALC) water. (Modified from Ayoob, S., Gupta, A.K., Bhakat, P.B. and Bhat, V.T., *Chem. Eng. J.*, 140, 6–14, 2008.)

The equation implies that OH^- is released from the ALC surface into the bulk phase, which is confirmed by the rise in pH during fluoride sorption in both natural and synthetic water and is more prominent at the initial stages. To experimentally quantify the process represented by the equations cited earlier and to understand spectroscopic changes in the adsorbent due to fluoride sorption, FTIR analysis was performed both before and after adsorption[3]

in synthetic and natural water. As shown in Figure 8.17, the FTIR spectrum of the samples presents no significant spectroscopic change due to fluoride sorption. The broad band corresponding to 3469 cm^{-1} (range of 3550–3200 cm^{-1}) represents O–H stretching vibrations, that at 1420 cm^{-1} represents Al–H stretching, and that at 1015 cm^{-1} represents the characteristic stretching bands of Al=O. The band at 570 cm^{-1} may be ascribed to the stretching of Al–OH, that at 662 cm^{-1} represents Si–H, and that at 872 cm^{-1} indicates Fe–O stretching.[3]

On closer examination, it has been observed that the intensity of many of the peaks shows variations after fluoride sorption. Before adsorption, in the virgin adsorbent (ALC), this peak height ratio was 1.0256, but after fluoride sorption, it was found to be 2.1348 in synthetic water (FALC) and 1.7613 in natural water (FALCN). This shows that the OH^- band at 3750 cm^{-1} is decreasing due to fluoride sorption, confirming the exchange between OH and F ions, and enhancing fluoride removal. In a similar way, the peak height ratio of OH^- band at 3469 cm^{-1} to that of Al–OH band at 561 cm^{-1} is also compared.

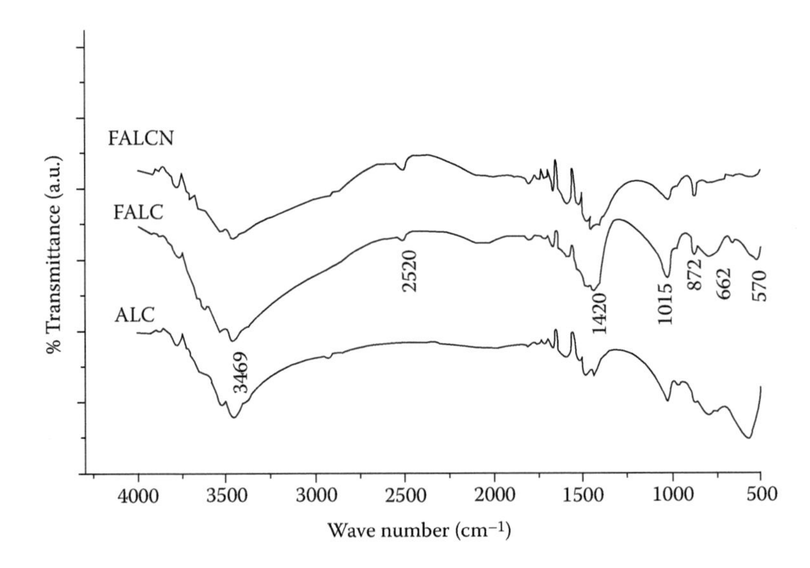

FIGURE 8.17
FTIR spectra of the adsorbent before sorption (ALC) and after sorption of fluoride in synthetic (FALC) and natural (FALCN) water. (Modified from Ayoob, S., Gupta, A.K., Bhakat, P.B. and Bhat, V.T., *Chem. Eng. J.*, 140, 6–14, 2008.)

> The ratio of the peak height was calculated as 0.99268 in ALC, whereas it is found to be 2.129 and 3.08 in FALC and FALCN, respectively. This clearly indicates that the OH ions of the Al–OH band are consumed in fluoride sorption, more prominently for the exchange reactions. Further, these observations are experimentally supported by the observed increase in pH (6.9 ± 0.4 to 11.7 ± 0.4) due to fluoride uptake, especially at the initial stages of sorption.[3]

So, the predominant mechanism of fluoride removal can be illustrated as shown in Figure 8.18.

> The role of sodium in groundwater deserves special mention, as the reduction in fluoride sorption at high pH may also be attributed to the sodium occupying sorption sites and altering the surface structure. Under basic conditions, surface hydronium ions can dissociate, enabling the aluminum to serve as a Lewis acid toward cations (such as sodium, calcium, or iron). This suggests that cations will be sorbed onto the alumina surface under basic conditions,[3]

involving the formation of an outer-sphere complex as follows:

$$AlOH + Na^+ = AlONa + H^+ \tag{8.57}$$

Before sodium sorption occurs, =Al–OH and =Al–O$^-$ sites are present in an alkaline solution. It is suggested that the sorption onto alumina (=ALONa) increases under alkaline conditions when the surface of the adsorbent is negatively charged.[42]

> This may lead to electrostatic repulsion between sorbed Na$^+$ ions and adjacent H$^+$ ions, causing some H$^+$ ions to break away from the surface; thus, fluoride sorption becomes reduced due to a reduction in available =AlOH sites. Due to the sodium occupying sorption sites, the concentrations of =Al–OH and =Al–O$^-$, in turn, reduce the total number of available sites for fluoride sorption. Since these reactions take place only when the surface charge of ALC is negative, this indicates that they occur during the late hours of sorption. This further supports the fact

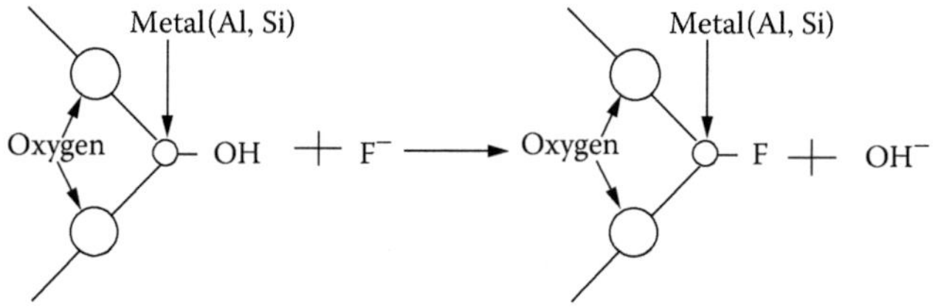

FIGURE 8.18
The ligand-exchange model of fluoride sorption onto ALC. (From Ayoob, S., Gupta, A.K., Bhakat, P.B. and Bhat, V.T., *Chem. Eng. J.*, 140, 6–14, 2008.)

that surface reactions are rate limiting. Thus, the fluoride sorption onto ALC may be viewed as an inner-sphere complexation that is predominated by a ligand-exchange process. The surface reactions leading to mixed surface complex precipitative reactions or scavenging reactions may also be involved.[3]

8.5.6 Isotherm Studies

The fitting of equilibrium sorption data to the Langmuir, Freundlich, and D–R isotherm models, for a wide range of fluoride concentrations (2.5–100 mg/L) in synthetic water, is shown in Figure 8.19. The visual inspection of the equilibrium sorption curves represented by the isotherm models indicate,

> A comparable uptake, of fluoride by ALC. An enhanced uptake of fluoride by ALC at a higher concentration is visible for LI and DRI models (with a tendency to tail off at very high concentrations), whereas FI provides an identical uptake pattern at all concentrations.[4]

> The shape of the fluoride isotherm data suggests that both FI and LI models would provide a better fit to the experimental data at all concentration ranges compared with DRI. It can be observed that (Table 8.2) LI fit the data better at all ranges of concentration with very high consistent values of R^2 (>0.995). The FI also makes identical fitting, though with slightly lesser R^2 values. However, the FI fitting becomes better towards higher concentration ranges, with a steady increase in the value of R^2 from 0.9738 to 0.9955.[4]

Thus, the sorption of fluoride by ALC derives significance in the concentration ranges studied, as it deviates from the notion that FI usually fits better at low concentrations and LI fits better at higher concentrations.[62]

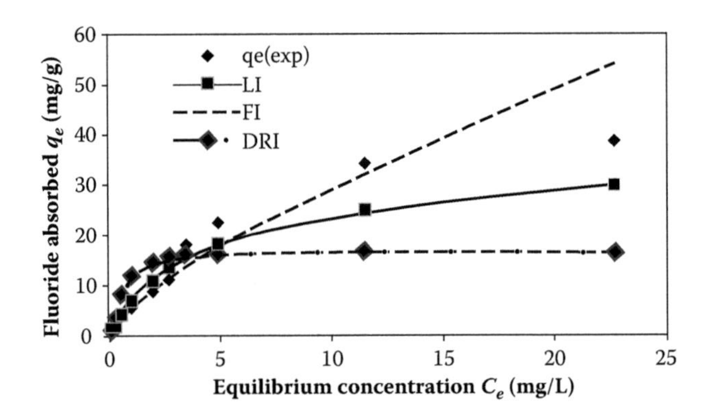

FIGURE 8.19
A comparison of the experimental data and various isotherm models in the concentration range of 2.5–100 mg/L of fluoride. (Modified from Ayoob, S. and Gupta, A.K. *J. Hazard. Mater.*, 152, 976–985, 2008.)

TABLE 8.2

Isotherm Model Parameters and R^2 Values in Natural and Synthetic Systems

Isotherm Model	Model Parameters and R^2	Range of Fluoride Concentrations (mg/L) Synthetic Samples at 300 K			
		2.5–20	2.5–40	2.5–80	2.5–100
LI	R^2	0.9952	0.9955	0.9954	0.9955
	b	0.329	0.2958	0.244	0.230
	q_{max} (mg/g)	24.57	27.17	32.57	34.36
FI	R^2	0.9738	0.9776	0.9843	0.9955
	$1/n$	0.7755	0.7939	0.7839	0.7248
	K_f	5.593	5.676	5.682	5.610
DRI	R^2	0.9588	0.8729	0.7821	0.7536
	K_{ad}	0.0791	0.0955	0.1135	0.1215
	Q_m (mg/g)	7.776	10.46	14.39	16.510

Source: Ayoob, S. and Gupta, A.K., *J. Hazard. Mater.*, 152, 976–985, 2008.

As cited earlier, this traditional approach of determining the isotherm parameters by linear regression of LI and FI models appears to give very good fits to the experimental data, as their respective R^2 values are very high (Table 8.2). Thus, purely from the comparison of R^2 values (being very close to unity), at all concentration ranges, the linearized Langmuir isotherm would be expected to provide a better fit to the experimental data than the linearized Freundlich isotherm.[4]

8.5.6.1 Effects of Temperature

The effects of temperature on all isotherm models in both natural and synthetic systems (with dose variations of ALC) are shown in Figure 8.20, with respective parameter values in Table 8.3. As can be seen, both LI and FI have shown a reasonably good fitting at all temperature ranges in both systems than DRI, though values of R^2 (Table 8.3) suggest that the Freundlich model fits better.[4]

8.5.7 Performance Evaluation of ALC in Natural and Synthetic Systems

Though the sorption profiles of fluoride removal were found to be identical in both synthetic and natural waters, ALC exhibited a reduced fluoride adsorption capacity in treating natural water compared with synthetic systems. The optimum dose requirement of ALC in natural water is five times more than that for synthetic water. The maximum monolayer saturation capacity of ALC showed identical reduction (with values of 10.214 mg/g in synthetic waters, and 0.9358 mg/g in natural waters).[2] Since the factors influencing the adsorption process and capacity of the adsorbent include pH, ionic strength, interference effects from other counter-ions, initial adsorbate concentrations, and temperature, these effects were investigated to elucidate the reasons for its reduced sorption capacity in natural water.

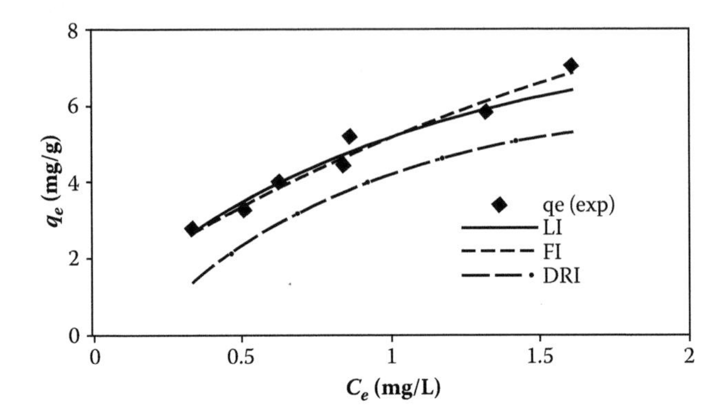

FIGURE 8.20
A comparison of the experimental data and various isotherm models in synthetic samples under dose variation study. (Modified from Ayoob, S. and Gupta, A.K., *J. Hazard. Mater.*, 152, 976–985, 2008.)

TABLE 8.3

Isotherm Model Parameters and R^2 Values in Natural and Synthetic Systems at Different Temperatures

Isotherm Model	Model Parameters and R^2	Dose Variation Study Temperature of Observations and Nature of Samples: 290 K Natural	290 K Synthetic	300 K Natural	300 K Synthetic	310 K Natural	310 K Synthetic
LI	R^2	0.8768	0.9877	0.9215	0.9599	0.9672	0.9918
	b	1.0336	0.8245	15.22	1.0471	10.696	1.3036
	q_{max} (mg/g)	1.0779	9.09	0.9358	10.215	1.164	12.658
FI	R^2	0.9275	0.9937	0.9364	0.9734	0.9391	0.9878
	$1/n$	0.3342	0.5569	0.1022	0.5959	0.1553	0.6287
	K_f	0.5589	3.980	0.825	5.1924	0.9939	7.5198
DRI	R^2	0.7440	0.9048	0.8826	0.9278	0.9463	0.4651
	K_{ad}	0.2046	0.0864	0.0136	0.1291	0.0171	0.0572
	Q_m (mg/g)	0.9157	6.9302	0.9157	6.1978	1.140	8.2137

Source: Ayoob, S. and Gupta, A. K., *J. Hazard. Mater.*, 152, 976–985, 2008.

8.5.7.1 Effect of pH, Ionic Strength, and Temperature

Since the sorption studies on synthetic water were also conducted at the same pH of natural water (6.9 ± 0.4), the reduction in the adsorption potential in the natural system could not be ascribed to the effect of pH. However, the sorptive responses of ALC were investigated at different values of pH to understand the mechanism and to ascertain its field use. As shown in Figure 8.14, the fluoride removal percentage was almost consistent in the pH range of 3–11.5 and was reduced thereafter. This could be readily explained as follows: When pH < pH_{ZPC}, the net surface

> charge on the solid surface of ALC is positive due to the adsorption of excess H^+, which favors adsorption due to coulombic attraction; whereas at pH > pH_{ZPC}, the net surface charge is negative due to desorption of H^+ and adsorption must compete with columbic repulsion.[2]

The observed reduction in fluoride adsorption above pH 11.5 may suggest that the strong negative surface charge developed at this pH may prevent fluoride from occupying available adsorption sites through columbic repulsion.[2] The effect of ionic strength on fluoride sorption onto ALC is shown in Figure 8.14. Experimental evidence suggests that anions that form inner-sphere complexes coordinate directly with the oxide surface without getting influenced by ionic strength.[60] Since the fluoride sorption is almost unaffected by the ionic strength, it can be surmised that the removal of fluoride occurs mainly through the formation of the inner-sphere surface complexation process.

> Since sorption studies on both systems were conducted at the same temperatures of 300 K, the reduction in the adsorption potential in natural water could not be ascribed to the effect of temperature. However, the temperature effects of sorption were evaluated in both systems within the range from 290 to 310 K to delineate the nature of sorption mechanisms in terms of the thermodynamic parameters: Gibbs free energy ($\Delta G°$), standard enthalpy ($\Delta H°$), and standard entropy changes ($\Delta S°$).[2]

The ln K versus $(1/T)$ plot is shown in Figure 8.13. The respective parameter values are illustrated in Table 8.4.

> The increase in fluoride sorption with temperature in both systems reflects the surface heterogeneity of ALC and its increased activity, which results in enhanced diffusion of fluoride ions into its pores. It would be expected that higher temperatures stimulate the surface reactivity of the bound oxides/hydroxides, which increases the sorption capacity of the system. The negative values of $\Delta G°$ confirm the spontaneity of sorption in both systems within the conditions applied. The higher negative value at elevated temperatures assures more energetically favorable adsorptions.

TABLE 8.4

Thermodynamic Parameters of Adsorption of Fluoride onto ALC in Natural and Synthetic Water

System	$\Delta H°$ (kJ/mol)	$\Delta S°$ (kJ/mol/K)	$\Delta G°$ (kJ/mol)		
			290 K	300 K	310 K
Synthetic Water	36.569	0.144	–5.191	–6.631	–8.071
Natural Water	83.810	0.301	–3.480	–6.490	–9.50

Source: Ayoob, S. and Gupta, A.K., *Chem. Eng. J.*, 150, 485–491, 2009.

> Further, the decrease in the magnitude with increasing temperature indicates more efficient sorption at elevated temperatures.[2]

The endothermic nature of the process in both systems is confirmed by the positive enthalpy values ($\Delta H°$). The positive value of entropy change ($\Delta S°$) reflects the affinity of ALC toward fluoride, which may also indicate some structural changes within the adsorbent.[2,63,64] Generally, the enhanced adsorption at elevated temperatures indicates that chemisorption is taking place in the system.[2]

8.5.7.2 Effects of Other Ions

The possible interferences from other ions that are commonly present in natural water were also investigated to examine their influences in the fluoride sorption onto ALC. Nitrates, chlorides, sulfates, and bicarbonates did not significantly affect the sorption process. The effect, of silicate was insignificant up to 25 mg/L; thereafter, it slightly reduced the fluoride removal and at 400 mg/L, the reduction was about 13%. The presence of calcium gradually enhanced the removal and at 400 mg/L, by around 7%. The presence of iron also did not affect the removal up to 10 mg/L, whereas the interference of phosphate was considerable. At a concentration of 4 mg/L, it reduces the removal by around 6% and at 8 mg/L, by more than 10%. The interference analyses suggest that high levels of salinity and hardness in water do not affect the fluoride removal performance of the adsorbent.

> Since fluoride-rich groundwaters are generally associated with high bicarbonate ions in alkaline environments, their applications turn out to be important. The interference of phosphate and silicate associated with fluoride sorption was already reported. The reduction in fluoride sorption in the presence of high silicates may be due to its scavenging of aluminum ions forming aluminosilicate solute species, especially in an alkaline environment. Also, silicic acid is known to inhibit the formation of aluminum hydroxide precipitates by replacing hydroxylated aluminum(III) ions. High silica concentrations may also result in its polymerization, resulting in an increase in negative surface charge. The reduction in the fluoride removal efficiency in the presence of phosphates may be due to the strong affinity of aluminum(III) for phosphate, thereby reducing the availability for fluoride uptake. The absence or negligible competitive effect produced from most of the ions indicates that fluoride is strongly adsorbed onto ALC. The interference pattern indicates that the presence of ion chlorides, nitrates, and bicarbonates may form outer-sphere complexes but sulfates and silicates form partial inner-sphere complexes with ALC.[2]

The total organic carbon (TOC) typically quantifies the amount of natural organic matter (NOM) concentrations in the natural water sources that are present as a result of adsorption onto aquifer solids or depositional history.[65] Aqueous NOM represents a wide range of structurally complex compounds that are derived from the chemical and biological degradation of plants and animals, composed mainly of humic substances (humic and fulvic acids), that are hydrophobic. Humic substances are complex mixtures containing both aromatic and aliphatic components with mainly carboxylic and phenolic functional groups; they were found to interfere with anionic adsorption[2,66–69] through stable metal complex formations.[70,71] In this study, additions of 5, 10, 20, and 40 mg/L of humic acid to fluoride samples (C_0 = 8.65 mg/L) reduced the percentage removal by 5.5%, 11.54%, 14.65%, and 18.03%, respectively.[2]

> As the structurally complex product of biomass decomposition, NOM molecules possess unique combinations of functional groups, including carboxylic, esteric, phenolic, quinone, amino, nitroso, sulfhydryl, hydroxyl, and other moieties, the majority of which are negatively charged at neutral pH. Along with this predominant anionic nature coupled with its high reactivity toward both metals and surfaces, NOM can compete with fluoride for sorption onto ALC. In sorptions involving natural groundwater, the adsorption capacity may depend on the accessibility of the organic molecules to the inner surface of the adsorbent. The small molecules can access micropores, and NOM can access mesopores of the adsorbent.[2,72]

This presorbed NOM reduces/destructs the sorption sites of the adsorbent. It was suggested that NOM readily forms both aqueous and surface inner-sphere complexes with cationic metals and metal oxides.[2,73]

> Aqueous NOM-metal complexes may, in turn, associate strongly with dissolved anions such as fluoride, presumably by metal-bridging mechanisms, diminishing the tendencies of fluoride ions to form surface complexes. This metal bridging appears to be a potential mechanism for reduced fluoride uptake in natural water. Though phosphates pose interferences to fluoride sorption in synthetic systems, since their respective concentrations in natural water are less, they may not be responsible for the reduced fluoride intake.[2]

However, the presence of silicate (Table 8.5), much above the average abundance level of 14 mg/L in groundwater,[6] may pose a slight interference.

> It is plausible that the presence of NOM represented by the high TOC value may play a role in the reduced capacity of the adsorbent in natural water. However, further studies are warranted to elucidate the reduced uptake in natural waters and the synergetic effects of various ions.[2]

8.5.8 Column Studies

The breakthrough curve developed from column studies basically portrays the dynamic sorptive responses of the adsorbent used in a continuous flow fixed bed. The shape of the breakthrough curve and the time for breakthrough

TABLE 8.5

Characteristics of Natural Groundwater (Collected from Baliasingh Patna, Kurda District, Orissa, India)

Characteristic Parameter	Quantitative Value
Fluoride (mg/L)	8.65
pH	6.9 ± 0.4
TDS (mg/L)	463
Acidity (mg/L)	1.5
Alkalinity (mg/L)	260
Chloride (mg/L)	165
Total hardness (mg/L)	145
Total organic carbon (mg/L)	59.08
Total phosphorous (mg/L)	0.032
Silicate as SiO_2 (mg/L)	39.22
Boron (mg/L)	0.33
Sodium (mg/L)	14.00
Potassium (mg/L)	2.00
Ammonia nitrogen (mg/L)	0.328
Salinity (PSS)	0.30[a,b]

Source: Ayoob, S. and Gupta, A. K. *Chem. Eng. J.*, 150, 485–491, 2009.
[a] Salinity is expressed in practical salinity scale (PSS).
[b] Minimum detection limit of salinity = 0.1 PSS.

appearance are predominant factors in determining the operation and dynamic response of an adsorption system. The general position of the breakthrough curve along the volume/time axis depends on the capacity of the column with respect to bed height, feed concentration, and flow rate. The point at which effluent fluoride concentration reaches 1 mg/L was taken as *breakthrough point* and that corresponding to 90% of influent concentration (C_t/C_0 = 0.90) was considered *point of exhaust*. Fixed bed column studies were undertaken in the up-flow mode to evaluate the performance of ALC in removing fluoride under varying operating conditions. During the sorption experiments, it was observed that the flow rate remained more or less constant, which indicated that the clogging of pores did not occur and, hence, the sorption sites of ALC were easily accessible through the inter particle pore network.

8.5.8.1 Effect of Process Parameters on Breakthrough

The breakthrough curves were obtained by varying the depths of the ALC bed (Z), flow rates (Q), and initial fluoride concentrations (C_0). A combined graph (Figure 8.21) shows the bed volumes of water treated at different stages. The respective service times, volumes of water treated, and the adsorption capacity of the columns under various process conditions are illustrated in Tables 8.6 and 8.7. It was observed that at lower C_t/C_0 ranges, the curves turn less steeper at higher bed depths under the same flow rates. Also, both breakthrough and exhaust times increase with the corresponding volumes

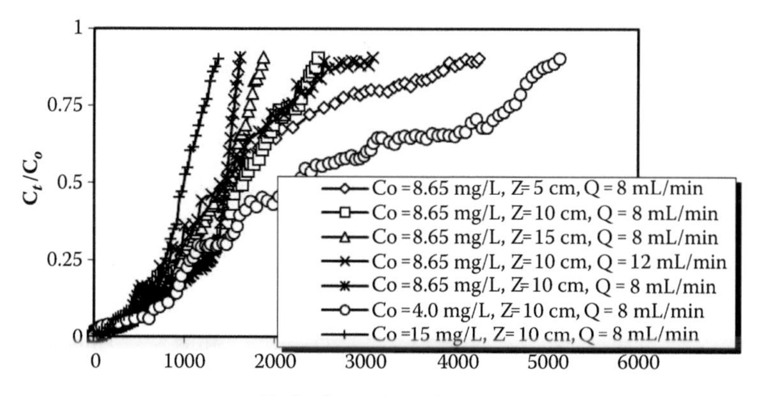

FIGURE 8.21
Experimental breakthrough curves of fluoride sorption onto ALC with bed volumes of water treated at different bed depths, flow rates, and initial fluoride concentrations in synthetic water. (From Ayoob, S. and Gupta, A.K., *Chem. Eng. J.*, 133, 273–281, 2007.)

TABLE 8.6

Sorption Data for Fixed Bed of ALC for Fluoride Sorption (Breakthrough) under Different Process Conditions in Synthetic Water

Initial Fluoride Concentration in mg/L (C_0)	ALC Bed Depth in cm (Z)	Flow Rate in mL/min (Q)	EBCT (min)	Time for Breakthrough in h (b_t)	Volume of Water Treated up to Point of Breakthrough in L (V_b)	Bed Volumes up to Point of Breakthrough	$q_{min,col}$ (mg/g)
8.65	5	8	4.5749	17	8.16	519.48	1.8466
8.65	10	8	9.1498	33	15.84	504.20	1.8021
8.65	15	8	13.7247	58	27.84	590.78	2.096
8.65	10	4	18.2995	84	20.16	641.71	2.270
8.65	10	12	6.0998	20	14.40	458.37	1.6355
4.00	10	8	9.1498	72	34.56	1100.08	1.722
15.00	10	8	9.1498	22	10.56	336.14	2.087

Source: Modified from Ayoob, S. and Gupta, A.K., *Chem. Eng. J.*, 133, 273–281, 2007.

of water treated, along with an increase in bed depths. The breakthrough volumes are found to increase around two times by doubling the depth of the bed from 5 to 10 cm; whereas they are more than three times for 15 cm.

> It is naturally expected that the availability of more adsorbent at higher bed depths offers more surface area and binding sites for sorption, resulting in enlarged mass transfer zones. However, the adsorption capacity of ALC at these different bed depths (5–15 cm) for a particular flow rate (8 mL/min) shows only marginal variation, indicating its consistency in affinity for fluoride sorption.[1]

TABLE 8.7

Sorption Data for Fixed Bed of ALC for Fluoride Sorption (Exhaust) under Different Process Conditions in Synthetic Water

Initial Fluoride Concentration in mg/L (C_0)	ALC Bed Depth in cm (Z)	Flow Rate in mL/min (Q)	EBCT (min)	Time for Exhaust in h (e_t)	Volume of Water Treated up to Exhaust in L (V_e)	Bed Volumes up to Exhaust	q_{col} (mg/g)
8.65	5	8	4.5749	139	66.72	4247.54	6.965
8.65	10	8	9.1498	162	77.76	2475.18	5.849
8.65	15	8	13.7247	184	88.32	1874.21	4.875
8.65	10	4	18.2995	212	50.88	1619.56	4.841
8.65	10	12	6.0998	134	96.48	3071.06	5.616
4.00	10	8	9.1498	336	161.28	5133.71	4.594
15.0	10	8	9.1498	90	43.20	1375.10	6.163

Source: Modified from Ayoob, S. and Gupta, A.K., *Chem. Eng. J.*, 133, 273–281, 2007.

> The breakthrough curves developed for the same bed height (10 cm) at higher flow rates appeared steeper, which may be due to the faster movement of the adsorption zone along the bed. The breakthrough capacity of the column ($q_{min,col}$) showed a consistent increase, though marginal, with a reduction in flow rates.

The total adsorption capacity of the column (q_{col}) also showed an identical response at 8–12 mL/min. An increase in the adsorption capacity at lower flow rates is usually expected due to better diffusivity of fluoride, resulting in enhanced sorption. The kinetic curve also reflects similar trends of sorption in the empty bed contact time (EBCT = volume of the bed/volumetric flow rate) ranges corresponding to these flow rates.[1]

It is also observed that the volumes of water treated at both breakthrough and exhaust increased more than two times, thus corresponding to a reduction in initial fluoride concentrations from 8.65 to 4.0 mg/L.

> Also, the service times of the column indicates that high fluoride concentrations aid quick saturation of the bed and enable faster movement of the adsorption zone. Since the higher concentration gradient between the solute in solution and the solute on the sorbent results in enhanced diffusion and sorption, the adsorption capacity of ALC is also found to increase at higher initial fluoride concentrations.[1]

The column performance of ALC renders an average adsorption potential of 5.896 mg/g at the point of exhaust. Though the column study suggests an adsorption potential of 5.896 mg/g, the maximum Langmuir monolayer adsorption capacity of ALC in batch studies was found to be much higher (10.215 mg/g).[1] Theoretically, the adsorption capacities from batch studies may not give accurate scale-up information about the column operation system[73]; whereas in fixed beds, the adsorption media have not been subjected to equilibrium sorption conditions and are, hence, not getting totally exhausted as in the batch system. Also, uneven flow patterns throughout the column may result in an incomplete exhaustion of bed, as cited earlier. However, in this study, the range of EBCT provided (~4.5–18.3 min) ensures around 75%–80% of the total removal observed in the batch system is represented by the kinetic curve.[1]

8.5.9 Application of Sorption Models

All the six models cited were applied to investigate the breakthrough behavior of fluoride sorption onto ALC in natural groundwater.

> The characteristic parameters of the models obtained by linear regression were used to predict the theoretical effluent fluoride concentrations. In the BDST model applications, the values of the maximum adsorption capacity parameter No. (Table 8.8) are found to decrease with increased bed depths, indicating that the adsorption zone is not moving with a constant speed along the column.[1] The adsorption rate constants (K), characterizing the rate of solute transfer from liquid to solid phase, were observed to increase with flow rate and initial fluoride concentrations, indicating the influence of external mass transfer on system kinetics. The higher values of *K* are advantageous, as they indicate that even a short bed will avoid breakthrough. Theoretically, the slope of the BDST line represents the time required for the adsorption zone to travel a unit length through the adsorbent bed.[1]

The predicted breakthrough times for 4 and 12 mL/min are found to be 80 and 23.33 h (using $b_t = 8.5Z - 5$ for 4 mL/min and $b_t = 2.833Z - 5$ for 12 mL/min), respectively, with corresponding exhaust times of 202.33 and 145.66 h (using $e_t = 8.5Z + 117.33$ for 4 mL/min and $e_t = 2.833Z + 117.33$ for 12 mL/min).[1] The time required for the adsorption zone to travel a unit length through

TABLE 8.8

Characteristic Parameters Predicted by BDST Model in Synthetic Water

Initial Fluoride Concentration (mg/L)	Bed Depth (cm)	Flow Rate (mL/min)	BDST Model			
			K (L/mg/h)	N_0 (mg/L)	R^2	APE
8.65	5	8	0.00482	18573.56	0.9048	38.853
8.65	15	8	0.00357	11794.47	0.9613	17.474
8.65	10	4	0.00251	12574.39	0.8916	22.057
8.65	10	12	0.00489	13653.96	0.9519	20.486
4.00	10	8	0.00323	11039.11	0.8794	27.440
15.0	10	8	0.00456	14760.46	0.9778	16.389

Source: Modified from Ayoob, S. and Gupta, A.K., *Chem. Eng. J.*, 133, 273–281, 2007.

the 10-cm adsorbent bed is as follows: 9.191, 4.25, and 2.4508 h for initial fluoride concentrations of 4, 8.65, and 15 mg/L, respectively. The predicted breakthrough times for 4 and 15 mg/L are 86.072 and 20.9814 h, respectively (using $b_t = 9.191Z - 5.838$ for 4 mg/L and $b_t = 2.4508Z - 3.5266$ for 15 mg/L) with corresponding exhaust times of 345.626 and 92.168 h[1] (using $e_t = 9.191Z + 253.326$ for 4 mg/L and $e_t = 2.4508Z + 67.66$ for 15 mg/L). A comparison of these predicted service times with the experimental service times indicates that the predicted service times are more than the corresponding experimental values (Table 8.9). Also, as per Equation 3.40, a curve between service times and bed heights at the 50% breakthrough point must result in a straight line passing through the origin if the process follows this model. In this study, the t_{50} values for 5-, 10-, and 15-cm bed depths are 50.24, 108.45, and 146.78 h, respectively. The plot of Z versus t_{50} offers a straight line plot ($R^2 = 0.9858$) but quite deviates from the origin (Figure 8.22). Generally, this failure is attributed to the complexity of the sorption process.[1] So, it is rational to believe that intra particle diffusion and external mass resistance are considerable and rate limiting in this complex

TABLE 8.9

Comparison of Experimental Service Times with Those Predicted by BDST Model under Different Process Conditions in Synthetic Water

Initial Fluoride Concentration in mg/L (C_0)	ALC Bed Depth in cm (Z)	Flow Rate in mL/min (Q)	Time for Breakthrough in Hours (b_t)		Time for Exhaust in Hours (e_t)	
			Experimental	Predicted	Experimental	Predicted
8.65	5	8	17	16.25	139	138.58
8.65	15	8	58	58.75	184	181.08
8.65	10	4	84	80	212	202.33
8.65	10	12	20	23.33	134	145.66
4.00	10	8	72	86.072	336	345.062
15.0	10	8	22	20.981	90	92.165

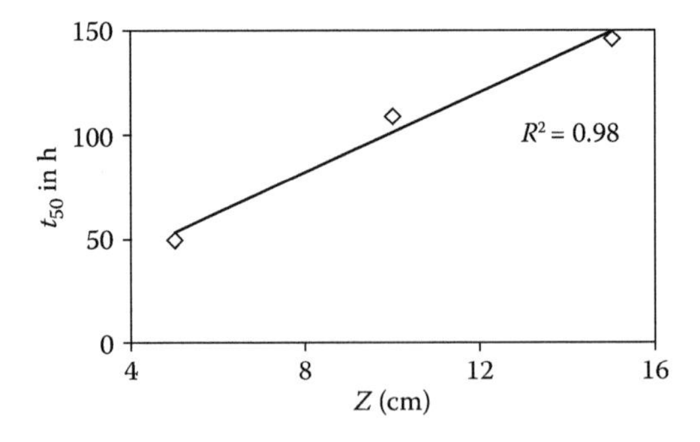

FIGURE 8.22
Bed depth versus time for 50% breakthrough plot of fluoride sorption onto ALC fixed bed in synthetic water.

fluoride sorption process and that their kinetics is controlled not alone by the surface chemical reactions between fluoride and ALC.[5]

> The linear regression of the Thomas model with the experimental fluoride sorption data also shows good correlations in most of the cases (Table 8.10). As shown, both the parameters q_{Th} and K_{Th} of the model were found to decrease with higher bed depths. For all conditions of bed depths, flow rates, and initial fluoride concentrations, the model predicted a marginally higher sorption capacity q_{Th} than experimental q_0 values. The rate constant (K_{YN}) of the Yoon–Nelson model decreased with an increase in bed depths but increased with flow rates and initial fluoride concentrations (Table 8.11). From the experimental results and data regression, it could be said that the Yoon–Nelson model provided a good correlation of the sorption of fluoride by ALC in most of the cases.[1]

The prediction of time for 50% breakthrough correlated well in higher flow rates and higher concentrations but differed much in other cases.[5]

TABLE 8.10

Characteristic Parameters Predicted by Thomas Model for Synthetic Water

Initial Fluoride Concentration (mg/L)	Bed Depth (cm)	Flow Rate (mL/min)	Thomas Model				
			K_{Th} (L/mg/h)	q_0 (mg/g)	q_{Th} (mg/g)	R^2	APE
8.65	5	8	0.0048	6.9652	7.9700	0.9048	38.853
8.65	15	8	0.0035	4.8747	5.0620	0.9613	17.474
8.65	10	4	0.0025	4.8413	5.3969	0.8916	22.057
8.65	10	12	0.0049	5.6160	5.8590	0.9519	20.486
4.00	10	8	0.0032	4.5940	4.7380	0.8794	27.440
15.0	10	8	0.0046	6.1630	6.3350	0.9778	16.389

Source: Modified from Ayoob, S. and Gupta, A.K., *Chem. Eng. J.*, 133, 273–281, 2007.

TABLE 8.11

Characteristic Parameters Predicted Yoon–Nelson Model for Synthetic Water

Initial Fluoride Concentration (mg/L)	Bed Depth (cm)	Flow Rate (mL/min)	Yoon–Nelson Model				
			K_{YN} (h^{-1})	τ (h)	τ_{cal} (h)	R^2	APE
8.65	5	8	0.0417	50.240	70.256	0.9048	38.853
8.65	15	8	0.0309	146.78	133.86	0.9613	17.474
8.65	10	4	0.0217	191.72	190.29	0.8916	22.057
8.65	10	12	0.0423	64.330	68.870	0.9519	20.486
4.00	10	8	0.0129	148.00	180.63	0.8794	27.440
15.0	10	8	0.0684	65.20	64.40	0.9778	16.389

Source: Modified from Ayoob, S. and Gupta, A.K., *Chem. Eng. J.*, 133, 273–281, 2007.

> Since the fluoride sorption by ALC follows the Freundlich isotherm model, its constant n was used for evaluating the parameters in Clark model. The model renders a good fit with comparatively higher R^2 and lower APE values in most of the experimental conditions (Table 8.12).[1]

The value of r is observed to increase with flow rates and initial fluoride concentrations but it decreases with bed depths. The magnitude of r increased more than two times, thus corresponding to the increase in flow rates from 4 to 12 mL/min. Similarly, the value of r increased more than two times when the initial fluoride concentration increased from 4 to 8.65 mg/L; whereas it was more than five times for 8.65–15 mg/L.[1]

The Wolborska sorption model was applied to experimental data for describing the initial part of the breakthrough curve ($C_t/C_0 \sim 0.5$). The model offers a reasonably good fitting with experimental data at lower flow rates and bed depths. The values of β and N_0 that are obtained from linear regression of the model are shown in Table 8.13. The values of the kinetic constant β were much influenced by flow rate and sharply increase with an increase in flow rates. With an increase in initial fluoride concentrations, they exhibit

TABLE 8.12

Characteristic Parameters Predicted by Clark Model for Synthetic Water

Initial Fluoride Concentration (mg/L)	Bed Depth (cm)	Flow Rate (mL/min)	Clark Model			
			ln A	r (h^{-1})	R^2	APE
8.65	5	8	1.9276	0.0362	0.9290	35.783
8.65	15	8	2.9132	0.0258	0.9635	15.676
8.65	10	4	2.8734	0.0177	0.8623	22.736
8.65	10	12	1.986	0.0374	0.9696	17.140
4.00	10	8	1.4816	0.0113	0.9022	25.456
15.0	10	8	3.109	0.0571	0.9694	18.636

Source: Modified from Ayoob, S. and Gupta, A.K., *Chem. Eng. J.*, 133, 273–281, 2007.

TABLE 8.13

Characteristic Parameters Predicted by Wolborska Model for Synthetic Water

Initial Fluoride Concentration (mg/L)	Bed Depth (cm)	Flow Rate (mL/min)	Wolborska Model			
			β (h^{-1})	N_0 (mg/L)	R^2	APE
8.65	5	8	116.907	14303.36	0.9424	23.179
8.65	15	8	40.532	13536.95	0.8948	28.891
8.65	10	4	28.818	15979.27	0.9319	18.441
8.65	10	12	79.155	13399.07	0.8654	29.990
4.00	10	8	49.105	9723.76	0.8939	24.480
15.0	10	8	63.013	17601.35	0.9364	20.274

TABLE 8.14

Characteristic Parameters Predicted by Bohart and Adams Model for Synthetic Water

Initial Fluoride Concentration (mg/L)	Bed Depth (cm)	Flow Rate (mL/min)	Bohart and Adams Model			
			K_{AB} (L/mg/h)	N_0 (mg/L)	R^2	APE
8.65	5	8	0.00817	14303.36	0.9424	23.179
8.65	15	8	0.00299	13536.95	0.8948	28.891
8.65	10	4	0.00180	15979.27	0.9319	18.441
8.65	10	12	0.00591	13399.07	0.8654	29.990
4.00	10	8	0.00505	9723.76	0.8939	24.480
15.0	10	8	0.00358	17601.35	0.9364	20.274

identical trends. For an increase in flow rate from 4 to 8 mL/min, the percentage increase in β is ~50% and further, for an increase from 8 to 12 mL/min, it is ~28%. The increased turbulences that developed within the bed at higher flow rates may reduce the film boundary layer surrounding the adsorbent, thereby increasing the β values. Practically, this results in faster breakthrough, considerably reducing the service time of columns, as is evidenced at higher flow rates and concentrations. This sorption behavior indicates that the system kinetics of fluoride sorption is dominated by external mass transfer, especially at initial breakthrough.[52] As expected, maximum adsorption capacity values (N_0) were reduced by increased flow rates from 4 to 12 mL/min. The Bohart and Adams model reflects similar features to the Wolborska sorption model in modeling the experimental data within the low range of breakthrough applied.[5] An identical increase in kinetic rates and adsorption capacities is also observed, as shown in Table 8.14.

8.5.9.1 Comparison of the Applied Models (Synthetic Water)

In general, the linear fittings of Hutchins BDST, Thomas and Yoon–Nelson models demonstrate that only the characteristic parameters associated with these models vary. However, all the three models will predict essentially the same C_t/C_0 values for a particular data set, and they are bound to give the same APE and R^2 values as illustrated earlier. But the prominent and unique characteristic features of the respective models, such as service time (Hutchins BDST model), adsorption capacity (Thomas model) and time for 50% breakthrough (Yoon–Nelson model), enable a further comparison to be made. The APE values on service time, adsorption capacity and time for a 50% breakthrough are 7.9952%, 6.1998% and 11.64%, respectively. So, the predictions by the Thomas model on its characteristic parameters turn more appropriate (as it renders the least error among the three) followed by the Hutchins BDST model and the Yoon–Nelson model.[1]

TABLE 8.15

Comparison of the Coefficient of Regression Values (R^2) Obtained from Linear Regression of Clark and Bohart and Adams Model (or Wolborska Model) with Experimental Data at Different Stages of Fluoride Sorption in Synthetic Water

Initial Fluoride Concentration (mg/L)	Bed Depth (cm)	Flow Rate (mL/min)	Breakthrough Point (Up to $C_t = 1$ mg/L)		50% Breakthrough ($C_t/C_0 = 0.5$)	
			CM	BAWM	CM	BAWM
8.65	5	8	0.9244	0.9353	0.9769	0.9424
8.65	15	8	0.9638	0.9527	0.9585	0.8948
8.65	10	4	0.9569	0.9464	0.9536	0.9319
8.65	10	12	0.9603	0.966	0.9351	0.8654
4.00	10	8	0.9446	0.9623	0.9394	0.8939
15.0	10	8	0.9662	0.9628	0.9552	0.9364

Note: CM, Clark model; BAWM, Bohart and Adams (or Wolborska model).

On comparing the R^2 and APE values of the Thomas model with those of the Clark model, it could be seen that the latter correlates marginally better. So, though the Thomas model could also describe the sorption process fairly well, the Clark model is deemed to be the best fit in terms of its slightly higher R^2 and lower APE values. The Bohart and Adams model and Wolborska model, plotted against the same axial settings of $\ln(C_t/C_0)$ against t, turn equivalent when k_{AB} becomes equal to β/N_0. As discussed earlier, both these models will also render similar APE and R^2 values and are equally good in correlating the sorption process in low breakthrough ranges. To have a meaningful comparison of the Clark model with the Bohart and Adams (or Wolborska) models, their correlations were also evaluated at lower breakthrough ranges (Table 8.15). It can be seen that the Clark model correlates better than the Bohart and Adams model (or Wolborska model) in all process conditions up to a 50% breakthrough; whereas both models are equally good in describing the process up to the point of breakthrough. So, in practical single-column applications or unit applications such as DDU, the Bohart and Adams model (or Wolborska model) can also be used, as it can describe the process up to the point of breakthrough fairly well. Thus, it can be concluded that since the Clark model could describe the process with the same vigor in all sorption ranges, it would be the most appropriate model to describe the sorption of fluoride onto ALC.[1,5]

8.5.9.2 Comparison of the Applied Models (Natural Water)

The features of the model are compared as in synthetic water studies, as discussed earlier. In case of fluoride sorption onto ALC in natural water, the APE values on service time (Hutchins BDST model), adsorption capacity (Thomas model), and time for 50% breakthrough (Yoon–Nelson model) are 20.57%, 8.29%, and 18.46%, respectively. So, the predictions by the Thomas model on the characteristic parameters turn more appropriate followed by the Yoon–Nelson

TABLE 8.16

Comparison of the Coefficient of Regression Values (R^2) Obtained from Linear Regression of Thomas and Bohart and Adams Model (or Wolborska Model) with Experimental Data at Different Stages of Fluoride Sorption in Natural Water

Initial Fluoride Concentration (mg/L)	Bed Depth (cm)	Flow Rate (mL/min)	Breakthrough Point (Up to C_t = 1 mg/L)		50% Breakthrough Point (C_t/C_0 = 0.5)	
			TM	BAWM	TM	BAWM
8.65	5	8	0.9063	0.8989	0.9745	0.9601
8.65	10	8	0.9375	0.9437	0.9634	0.9775
8.65	15	8	0.8731	0.8812	0.9391	0.9478
8.65	10	4	0.9507	0.9588	0.9405	0.9500
8.65	10	12	0.9145	0.9161	0.9348	0.9658

and Hutchins BDST models. It was observed that the Clark model fails to fit the sorption data, rendering poor R^2 and high APE values. Thus, it appears that the Bohart and Adams model (or Wolborska model) could describe the sorption process fairly well up to a 50% breakthrough, whereas the Thomas model could describe the process up to exhaust. Further, to differentiate between these models, their applications at different stages of breakthrough are compared. The correlation of these models with the experimental fluoride sorption data at these stages is illustrated in Table 8.16. It becomes very obvious that the Thomas model could describe the sorption of fluoride onto ALC at all these stages of the sorption process, whereas the Bohart and Adams model (or Wolborska model) could be applied only up to low breakthrough ranges. Practically, in single-column applications or unit applications such as DDU, both models may describe the process with the same vigor. However, in pilot plants and field applications involving series of columns, the Thomas model would be the most appropriate. The conventional practice of comparing the sorption system responses with predicted values of the model involves plotting the respective curves against the same axial settings, usually C_t/C_0 versus $t^{1.5}$ (Figure 8.23).

8.5.10 Fluoride Desorption Studies

The desorption properties of the adsorbent ALC were tested in fixed bed studies by using an eludent of 10% NaOH in synthetic and natural systems. Exhausted columns of 5-cm ALC media were regenerated after the single run of fluoride sorption. The desorption profiles of ALC in both synthetic and natural water are shown in Figure 8.24. The 5-cm ALC bed sorbed 6.965 mg/g of fluoride, depicted during sorption studies conducted on synthetic water. The concentration fluoride desorbed in the first bed volume was 2350 mg/L, which got reduced thereafter and remained almost constant at around 4.2 mg/L after 20 bed volumes. On calculation, the total fluoride desorbed was found to be 3.729 mg/g only given that the percentage

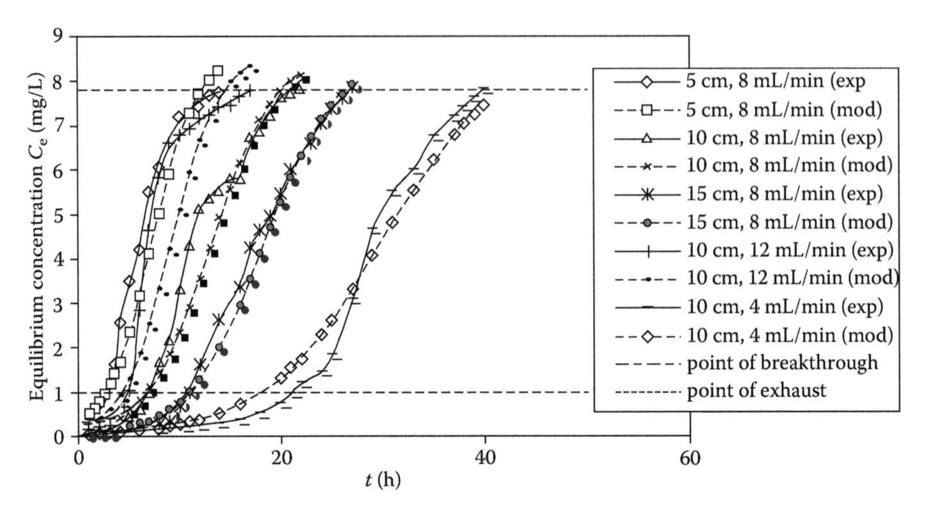

FIGURE 8.23
Comparison of experimental breakthrough profiles against theoretical values predicted by Thomas model under different process conditions in natural water. (From Ayoob, S., Gupta, A.K. and Basheer, A.B., *J. Urban Environ. Eng.*, 3, 17–22, 2009).

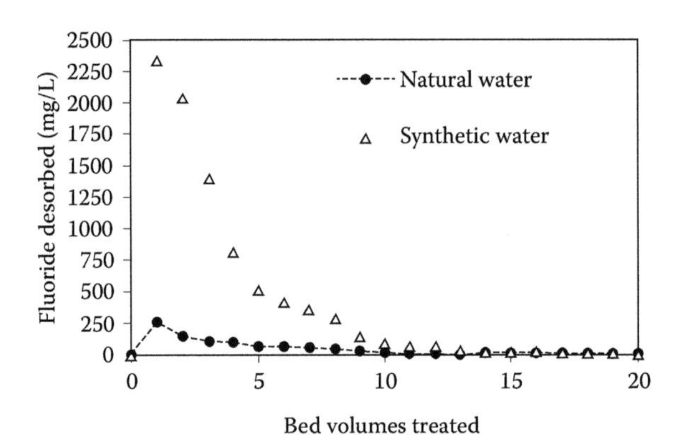

FIGURE 8.24
Fluoride desorption profile of ALC during regeneration with 10% NaOH in column studies.

of fluoride desorbed was 53.54%. Almost identical desorption profiles were observed for the ALC bed that sorbed 0.7503 mg/g of fluoride from natural water. Though the fluoride desorbed in the first bed volume was 250 mg/L, it was 2.15 mg/L after 20 bed volumes. The total fluoride desorbed was found to be 0.3571 mg/g, rendering the percentage of fluoride desorbed as 47.60%. In both systems, even after 20 bed volumes, the concentration of fluoride in the effluent remained high, indicating a very slow desorption profile. The poor desorptional characteristics may indicate a chemisorptive mechanism through a strong chemical interaction between fluoride and ALC.

This may indicate the irreversibility of fluoride sorption, which further suggests that ion exchange is not the only sorption mechanism involved in fluoride removal. It is also plausible that the fluoride ions that initially attached to ALC by ligand- or ion-exchange mechanisms may become more firmly bound by chemisorption. The chemical compounds formed due to sorption as indicated by the XRD, coupled with these poor desorption characteristics of ALC, further demonstrate that such transformations might have taken place.[3]

8.6 Summary of the Case Study

The world is in a water crisis. The presence of geogenic pollutants such as fluoride in groundwater exacerbates the gravity of this issue, thereby posing hindrance to the global onward march of humanity toward a *water secure world*. Of late, fluoride in drinking water has been wreaking havoc in more than 35 nations, forcing hundreds of millions of people to live in the shadows of fluorosis. Blame it on different forms of fluorosis, in many countries such as India and China, its excess presence in water is prioritized as a major impediment to the sustainable drinking water supply. As a result, defluoridation of drinking water has been regarded as one of the key areas of attention among the global water community, triggering global research. Since majority of the affected people worldwide live in small communities, more emphasis should be given to developing treatment technologies tailored for such habitations. Adsorption is an important technique that is the most widely used for excess fluoride removal in such habitats of the developing world. Though many adsorbents had been developed for defluoridation, only very few of them such as activated alumina were found to be useful in field applications. So, the development of economically viable and socially acceptable adsorbents with adequate fluoride removal potential has become a necessity. The adsorbent ALC developed in this direction was tested for its applicability as an adsorbent for defluoridation. The experimental investigations comprised a series of batch and continuous flow experiments performed under varying operating conditions. The results of such experimental observations and major findings of the research can be summarized as follows:

- Batch study results demonstrate the feasibility of ALC in removing fluoride from aqueous systems. It offers comparable defluoridation potential in both natural and synthetic systems. The kinetic curve of fluoride sorption exhibited a biphasic uptake with a rapid phase in the initial few minutes followed by a slow phase in both the systems. The slow phase, which governs the rate limiting, has been generally ascribed to diffusion into micropores, retention on sites of lower reactivity, and could also be due to surface nucleation/precipitation.

- At optimum conditions, an ALC dose of 10 g/L could remove 94.54% of fluoride from natural water with an initial fluoride concentration of 8.65 mg/L, whereas a lesser dose of 2 g/L could remove 92.76% fluoride in synthetic waters of the same concentration. The fluoride removal percentage decreased from 96.76% to 89.56% with an increase in initial fluoride concentrations from 2.5 to 50 mg/L; whereas, the adsorption capacity increased from 1.2 to 22.39 mg/g. This increase in adsorption capacity in the synthetic system at higher fluoride concentrations, signifying heterogeneous sorptive surfaces, may be ascribed to high intramolecular competitiveness to occupy the unsaturated lower energetic surface sites.
- The XRD results reveal multiple oxide surface sites, which are suggestive of sorption occurring on heterogeneous binding sites. This is further confirmed by an increase in sorption density with higher fluoride concentrations. Further, the variation in the affinity of ALC to different adsorbent doses indicates a surface-bound sorption. Since the adsorbent is a composite, its surface may be composed of sites with a spectrum of binding energies. It is postulated that at low adsorbent dosages all types of sites are entirely exposed to sorption, and the surface may become saturated faster. However, at higher adsorbent dosages, the availability of higher energy sites may decrease and a larger fraction of lower energy sites may become occupied. This results in an overall decrease in binding energy of the surface, and a reversible type of process may exist between the fluoride ions attached to low energy sites and those present in bulk solution. This may be the reason for the observed increase in uptake with an increase in adsorbent dosages up to a certain stage.
- The fitting of the kinetic data in both systems demonstrates that the dynamics of sorption could be well described by the pseudo-second-order model, which is indicative of rate-limiting chemisorption. The reasonable fitting of the kinetic data to the pseudo-first-order model, intra particle diffusion model, and Elovich model shows the possibilities of a diffusion-controlled yet complex sorption mechanism. The values of activation energy in natural and synthetic systems further support the role of diffusion in the sorption process.
- The experimental evidence pertaining to the response of the system to pH, inert electrolyte concentration, and desorption pattern suggests that the slow sorption phase in fluoride removal is not the result of diffusion (neither inter particle nor intra particle). Further, the rate limiting is due either to the heterogeneity of the surface site-binding energy or to other reactions controlling fluoride removal from the solution.
- The analysis of FTIR data together with an observed increase in pH during sorption suggests that a ligand-exchange mechanism is prominent in the removal process. The poor desorptional characteristics

of ALC indicate the irreversibility of fluoride sorption, which may further suggest that ion exchange is not the only sorption mechanism involved. The insensitivity of inert electrolyte concentrations to fluoride uptake suggests the possibilities of fluoride forming inner-sphere complexes with ALC. It is plausible that the fluoride ions initially attached to ALC by ligand- or ion-exchange mechanisms may become more firmly bound by chemisorption. The chemical compounds formed due to sorption (as indicated by the XRD) coupled with poor desorption characteristics of ALC further demonstrate that such transformations are plausible. Overall, the removal of fluoride may be viewed as a sorption process in which fluoride is *attached* to ALC through an inner-sphere complex formation through strong bonds (adsorption), diffusion into the crystal structure (absorption), and various surface precipitation reactions.

- The equilibrium data of fluoride sorption onto ALC conducted at different concentration ranges of fluoride could fit better with the Langmuir isotherm model than the Freundlich and DR isotherm models. At all concentration ranges tested, LI offered an excellent fit with very high consistent values of R^2 (>0.995). The maximum saturated monolayer capacity of ALC in synthetic water recorded an increase from 24.57 to 34.36 mg/g, corresponding to an increase in the initial fluoride concentration ranges from 2.5 to 20 mg/L and from 2.5 to 100 mg/L. The dose variation study rendered the maximum saturated monolayer capacity of ALC in synthetic water as 10.215 mg/g and in natural water as 0.9358 mg/g. So, ALC exhibited a reduction in adsorption capacity in treating natural groundwater compared with synthetic water.
- The adsorption capacity parameters of all isotherm models recorded an increase at high temperatures, indicating that the sorption of fluoride by ALC is endothermic, which is characterized by a chemisorption mechanism. The sorption of fluoride onto ALC was almost consistent in the pH range of 3–11.5 and was reduced thereafter. Also, the fluoride removal process was unaffected by ionic strength, indicating that the removal of fluoride occurs mainly through the formation of the inner-sphere surface complexation process.
- The endothermic nature of the process in both systems is further confirmed by the negative enthalpy values. The increase in fluoride sorption with an increase in temperature in both systems reflects the surface heterogeneity of ALC and its increased activity, which results in enhanced diffusion of fluoride ions into its pores. It would be expected that higher temperatures would stimulate the surface reactivity of the bound oxides/hydroxides, which increases the sorption capacity of the system. The positive value of entropy change reflects the affinity of ALC toward fluoride, which may also indicate

some structural changes within the adsorbent. The negative values of free energy confirm the spontaneity of sorption in both systems.

- The presence of nitrates, chlorides, sulfates, and bicarbonates did not significantly affect the sorption process. This may suggest that high levels of salinity and hardness in water did not affect the fluoride removal performance of ALC. Since fluoride-rich groundwaters are generally associated with high bicarbonate ions in alkaline environments, this may enhance the scope of ALC applications. The effect of silicate turns significant only after 25 mg/L, and at 400 mg/L it reduces fluoride removal by around 13%. The presence of iron also did not affect the removal up to 10 mg/L. At a concentration of 4 mg/L, the presence of phosphate reduces fluoride removal by around 6% and at 8 mg/L, by more than 10%. The presence of 40 mg/L of humic acid reduces fluoride removal by 18.03%. The reduction of fluoride sorption in the presence of high silicates may be due to its scavenging of aluminum ions, forming aluminosilicate solute species in the alkaline environment, and due to its polymerization, thereby resulting in an increase in negative surface charge. The reduction in fluoride removal efficiency in the presence of phosphates may be due to the strong affinity of aluminum for phosphate, reducing its availability for fluoride uptake. The interference pattern indicates that the presence of ion chlorides, nitrates, and bicarbonates may form outer-sphere complexes but sulfates and silicates form partial inner-sphere complexes with ALC. Aqueous NOM-metal complexes may, in turn, associate strongly with dissolved anions such as fluoride, presumably by metal-bridging mechanisms, diminishing the tendencies of fluoride ions to form surface complexes. This metal bridging appears to be a potential mechanism for reduced fluoride uptake in natural water. However, further experiments on groundwater samples from different locations and possessing a different chemistry are needed to properly elucidate this mechanism.
- The performance of ALC in column studies in synthetic water rendered an average adsorption potential of 5.896 mg/g at the point of exhaust. Also, the adsorption capacity of ALC at different bed depths for a particular flow rate showed only marginal variation. The breakthrough and total capacities of the column showed a consistent increase, though marginal, with a reduction in flow rates. The volumes of water treated at breakthrough and exhaust were reduced considerably with an increase in initial fluoride concentrations. High fluoride concentrations may aid quick saturation of the bed and enable faster movement of the adsorption zone. As the higher concentration gradient between the solute in solution and solute on the sorbent results in enhanced diffusion and sorption, the adsorption capacity of ALC is also found to increase at higher initial fluoride concentrations.

- Though the BDST model rendered reasonable fitting with the experimental data, the model fails to make fair predictions of service times in most of the cases. Further, the predicted service times are generally more than the corresponding experimental values. Also, the curve between service times and bed heights at a 50% breakthrough point was found to not pass through the origin, indicating its failure in accurately describing the sorption of fluoride onto ALC. This failure may be attributed to the complexity of the sorption process. So, it is plausible that the intra particle diffusion and external mass resistance are considerable and rate limiting in this complex fluoride sorption process and that its kinetics is controlled not alone by the surface chemical reactions between fluoride and ALC.
- The linear regression of the Thomas model with the experimental fluoride sorption data showed very good correlations in most of the cases. Moreover, for all conditions of bed depths, flow rates, and initial fluoride concentrations, the model predicted the sorption capacity fairly well. Though the Yoon–Nelson model provided good correlations, the prediction of time for a 50% breakthrough was inaccurate in most of the cases. The Clark model rendered a good fit with comparatively higher R^2 and lower APE values in most of the experimental conditions. The Bohart and Adams model reflected similar features to the Wolborska sorption model in modeling the experimental data within the low range of the breakthrough applied. Though the Thomas model could also describe the sorption process fairly well, the Clark model is deemed to be the best fit in terms of slightly higher R^2 values and lower APE values. It was also observed that the Clark model correlated better than the Bohart and Adams model (or Wolborska model) at all process conditions up to a 50% breakthrough; whereas both models were equally good in describing the process up to breakthrough.
- The performance of ALC in column studies with natural water rendered an average adsorption potential of 0.7064 mg/g at the point of exhaust. Also, as observed in synthetic water, the adsorption capacity of ALC at different bed depths for a particular flow rate showed only marginal variation. The breakthrough and total capacities of the column showed a consistent increase, though marginal, with a reduction in flow rates, thus indicating its consistency in affinity for fluoride sorption. The breakthrough curves appear steeper at higher flow rates; this may be due to faster movement of the adsorption zone along the bed, aiding its quick saturation. The observed reduction in column adsorption capacity in both natural and synthetic systems compared with its corresponding batch performance is reasonable, as batch studies may not give accurate scale-up information about the column operation system. In fixed beds, the adsorption media have not been subjected to equilibrium sorption conditions

and, hence, are not getting totally exhausted as in the batch sorption system. Also, uneven flow patterns throughout the column may result in an incomplete exhaustion of bed.

- The BDST model fairly predicted the service times of the column in natural water up to breakthrough, but it failed to correlate exhaust times in a better way. So, the model could fit well only with initial portions of the experimental breakthrough curve. Also, as in synthetic water, the curve between service times and bed heights at a 50% breakthrough point was found to not pass through the origin, indicating its failure in accurately describing the sorption process. This failure may be attributed to the complexity of the sorption process, indicating that the kinetics of the sorption process in natural water is controlled not alone by the surface chemical reactions between fluoride and ALC. The linear regression of the Thomas model with the experimental fluoride sorption data in natural water also showed very good correlations in most of the cases. Moreover, for all conditions of bed depths, flow rates and initial fluoride concentrations, the model predicted the sorption capacity fairly well. Though the Yoon–Nelson model also provided good correlations, the prediction of time for a 50% breakthrough was inaccurate in most of the cases. However, Clark model rendered a poor fit with the fluoride sorption data with comparatively low R^2 and higher APE values in all ranges of experimental conditions. The Bohart and Adams model reflected similar features to Wolborska sorption models in modeling the experimental data within the low range of breakthrough applied. Altogether, the Thomas model was demonstrated successful in describing the sorption of fluoride onto ALC at all stages of the sorption process, whereas the Bohart and Adams model (or Wolborska model) could be applied only up to low breakthrough ranges.

8.7 Conclusions of the Case Study

The synthesis of the ALC sorptive system for defluoridation of water has been convincingly demonstrated successful, as it showed comparable and consistent fluoride sorption potential in both batch and column applications. The capacity of the adsorbent in bringing the fluoride concentrations within the permissible limits over a wide range of pH without sacrificing its potential may be useful in treating natural fluoride-rich groundwaters. The removal may be viewed as a complex sorption process, occurring via multiple mechanisms, including electrostatic attraction, anion and ligand exchange, and other chemical reactions. The consistency in fluoride sorption over a wide pH range, availability and adaptability in the local environment, and comparable sorption potential of ALC in both synthetic and natural systems deserve merit. This study describes in detail the systematic procedures and protocols to be adopted in carrying out a study on the adsorptive removal of fluoride. The detailed investigations performed through batch and column studies convincingly demonstrate the feasibility of using ALC for fluoride removal in household and community applications in the endemic areas.

References

1. Ayoob, S. and Gupta, A.K. (2007). Sorptive response profile of an adsorbent in the defluoridation of drinking water. *Chem. Eng. J.*, 133, 273–281.
2. Ayoob, S. and Gupta, A.K. (2009). Performance evaluation of alumina cement granules in removing fluoride from natural and synthetic waters. *Chem. Eng. J.*, 150, 485–491.
3. Ayoob, S., Gupta, A. K., Bhakat, P.B. and Bhat, V.T., (2008). Investigations on the kinetics and mechanisms of sorptive removal of fluoride from water using alumina cement granules. *Chem. Eng. J.*, 140, 6–14.
4. Ayoob S. and Gupta A.K. (2008). Insights into isotherm making in the sorptive removal of fluoride from drinking water. *J. Hazard. Mater.*, 152, 976–985.
5. Ayoob, S., Gupta, A.K. and Basheer, A.B. (2009). A fixed bed sorption system for defluoridation of ground water. *J. Urban Environ. Eng.*, 3, 17–22.
6. APHA. (1998). *American Public Health Association, Standard Methods for the Examination of Water and Wastewater*, 20th edn. Washington, DC: American Public Health Association.
7. Das, D.P., Das, J. and Parida, K. (2003). Physicochemical characterization and adsorption behavior of calcined Zn/Al hydrotalcite-like compound (HTlc) towards removal of fluoride from aqueous solution. *J. Colloid Interface Sci.*, 261, 213–220.
8. UNICEF. (1999). *UNICEF Water Front, Fluoride in Water: An Overview*, p. 11. New York: UNICEF programme division, WES section, three United Nations plazas.
9. WHO. (2002). Fluorides, Environmental Health Criteria Number, 227. Geneva, Switzerland: World Health Organization.
10. Ayoob, S., Gupta, A. K., Bhakat, P.B. (2007). Analysis of breakthrough developments and modeling of fixed bed adsorption system for As(V) removal from water by modified calcined bauxite (MCB). *Sep. Purif. Technol.*, 52, 430–438.
11. Ho, Y.S. (2006). Review of second order models for adsorption systems. *J. Hazard. Mater.*, 136, 681–689.
12. Ho, Y.S., McKay, G. (1998). A comparison of chemisorption kinetic models applied to pollutant removal on various sorbents. *Trans. IChemE*, 76, 332–340.
13. Lagergren, S. (1898). About theory of so-called adsorption of soluble substances. *K. Svenska Ventenskapsakad Handlingar*, 24, 1–39.
14. Aharoni, C. and Sparks, D.L. (1991). Kinetics of soil chemical reactions: A theoretical treatment. In: Sparks, D.L. and Suarez, D.L., (Eds.), *Rates of Soil Chemical Processes*, SSSA special publication no. 27, pp. 1–18. Madison, WI: Soil Science Society of America.

15. Crini, G., Peindy, H.N., Gimbert, F. and Robert, C. (2007). Removal of C.I. Basic Green 4 (Malachite Green) from aqueous solutions by adsorption using cyclodextrin-based adsorbent: Kinetic and equilibrium studies. *Sep. Purif. Technol.*, 53, 97–110.
16. Maliyekkal, S.M., Sharma, A.K. and Philip, L. (2006). Manganese-oxide-coated alumina: A promising sorbent for defluoridation of water. *Water Res.*, 40, 3497–3506.
17. Ghorai, S. and Pant, K.K. (2005). Equilibrium, kinetics and breakthrough studies for adsorption of fluoride on activated alumina. *Sep. Purif. Technol.*, 42, 265–271.
18. Weber, W.J. and Morris, J.C. (1963). Kinetics of adsorption on carbon from solution. *J. Sanit. Eng. Div. ASCE*, 89, 31–39.
19. Zhang, J. and Stanforth, R. (2005). Slow adsorption reaction between arsenic species and goethite (α-FeOOH): Diffusion or heterogeneous surface reaction control. *Langmuir*, 21, 2895–2901.
20. Hingston, F.J. (1981). A review of anion adsorption. In: Anderson, M.A. and Rubin, A. J., (Eds.), *Adsorption of Inorganics at the Solid-Liquid Interface*, p. 67. Ann Arbor, MI: Ann Arbor Science.
21. Aharoni, C., Sparks, D.L., Levinson, S. and Revina, I. (1991). Kinetics of soil chemical reactions: Relationships between empirical equations and diffusion models. *Soil Sci. Soc. Am. J.*, 55, 1307–1312.
22. Rudzinksi, W. and Panczyk, P. (1998). In: Schwarz, J.A. and Contescu, C.I., (Eds)., *Surfaces of Nanoparticles and Porous Materials*, p. 355. New York: Dekker.
23. Sparks, D.L. (2005). Sorption-desorption, kinetics. In: Hillel, D, Hatfield, J.L., Powlson, D.S., Rosenweig, C., Scow, K.M., Singer M.J. and Sparks, D.L. (Eds.), *Encyclopedia of Soils in the Environment*. pp. 556–561. Oxford, UK: Elsevier Ltd.
24. Adamson, A.W. and Gast, A.P. (1997). *Physical Chemistry of Surfaces*, 6th edn. New York: Wiley-Interscience.
25. Langmuir, I. (1916). The constitution and fundamental properties of solids and liquids. *J. Am. Chem. Soc.*, 38, 2221–2295.
26. Freundlich, H.M.F. (1906). Über die adsorption in lösungen. *Z. Phys. Chem.*, 57, 385–470.
27. Dubinin, M.M. (1960). The potential theory of adsorption of gases and vapors for adsorbents with energetically non-uniform surface. *Chem. Rev.* 60, 235–266.
28. Ceyhan, O. and Baybas, D. (2001). Adsorption of some textile dyes by hexadecyltrimethylammonium bentonite. *Turk. J. Chem.*, 25, 193–200.
29. Wolkenstein, T. (1991). *Electronic Processes on Semiconductor Surfaces during, Chemisorption*. New York: Consultants Bureau.
30. Kinnilburgh, D.G. (1986). General purpose adsorption isotherms. *Environ. Sci. Technol.*, 20, 895–904.
31. Qadeer, R., Hanif, J., Khan, M. and Saleem, M. (1995). Uptake of uranium ions by molecular sieve. *Radiochim. Acta*, 68, 197–201.
32. Ozcan, A., Oncu, E.M. and Ozcan, A.S. (2006). Kinetics, isotherm and thermodynamic studies of adsorption of Acid Blue 193 from aqueous solutions onto natural sepiolite. *Colloids Surf. A: Physicochem. Eng. Aspects*, 277, 90–97.
33. Kasprzyk-Hordern, B. (2004). Chemistry of alumina, reactions in aqueous solutions and its application in water treatment. *Adv. Colloid Interface Sci.*, 110, 19–48.
34. Weber, W.J. Jr. (1972). *Physicochemical Processes for Water Quality Control*. New York: A Wiley-Inter science Publication, John Wiley and Sons.

35. Kumar, K.V. and Sivanesan, S. (2006). Pseudo second order kinetics and pseudo isotherms for malachite green onto activated carbon: Comparison of linear and non-linear regression methods. *J. Hazard. Mater.*, 136, 721–726.
36. Hao, J.O. and Huang, C.P. (1986). Adsorption characteristics of fluoride onto hydrous alumina. *J. Environ. Eng. (ASCE)*, 112, 1054–1069.
37. Ayoob, S., Gupta, A.K and Bhat, V.T. (2008). A conceptual overview on sustainable technologies for the defluoridation of drinking water. *Crit. Rev. Environ. Sci. Technol.*, 38, 401–470.
38. Onyango, M.S., Kojima, Y., Aoyi, O., Bernardo, E.C. and Matsuda, H. (2004). Adsorption equilibrium modeling and solution chemistry dependence of fluoride removal from water by trivalent-cation-exchanged zeolite F-9. *J. Colloid Interface Sci.*, 279, 341–350.
39. Das, N., Pattanaik, P. and Das, R. (2005). Defluoridation of drinking water using activated titanium rich bauxite. *J. Colloid Interface Sci.*, 292, 1–10.
40. Tripathy, S.S., Bersillon, J.L. and Gopal, K. (2006). Removal of fluoride from drinking water by adsorption onto alum-impregnated activated alumina. *Sep. Purif. Technol.*, 50, 310–317.
41. Handa, B.K. (1975). Geochemistry and genesis of fluoride-containing ground waters in India. *Ground Water*, 13, 278–281.
42. Thomas, W.J. and Crittenden, B.D (1998). *Adsorption Technology and Design*. Oxford, UK: Reed Educational and Professional Publishing.
43. Schiewer, S. and Volesky, B. (1997). Ionic strength and electrostatic effects in biosorption of divalent metal ions and protons. *Environ. Sci. Technol.*, 31, 2478–2485.
44. Gao, Y. and Mucci, A. (2001). Acid base reactions, phosphate and arsenate complexation, and their competitive adsorption at the surface of goethite in 0.7 M NaCl solution. *Geochim. Cosmochim. Acta*, 65, 2361–2378.
45. Kundu, S. and Gupta, A.K. (2007). As(III) removal from aqueous medium in fixed bed using iron oxide-coated cement (IOCC): Experimental and modeling studies. *Chem. Eng. J.*, 129, 123–131.
46. Hutchins, R.A. (1973). New simplified design of activated carbon system. *Am. J. Chem. Eng.*, 80, 133–138.
47. Thomas, H.C. (1944). Heterogeneous ion exchange in a flowing system. *J. Am. Chem. Soc.*, 66, 1664–1666.
48. Yoon, Y.H. and Nelson, J.H. (1984). Application of gas adsorption kinetics. I. A theoretical model for respirator cartridge service life. *Am. Ind. Hyg. Assoc. J.*, 45, 509–516.
49. Clark, R.M. (1987). Evaluating the cost and performance of field-scale granular activated carbon systems. *Environ. Sci. Technol.*, 21, 573–580.
50. Wolborska, A. (1989). Adsorption on activated carbon of p-nitrophenol from aqueous solution. *Water Res.*, 23, 85–91.
51. Bohart, G.S and Adams, E.Q. (1920). Some aspects of the behavior of charcoal with respect to chlorine. *J. Am. Chem. Soc.*, 42, 523–544.
52. Aksu, Z. and Gönen, F. (2004). Biosorption of phenol by immobilized activated sludge in a continuous packed bed: Prediction of breakthrough curves. *Process Biochem.*, 39, 599–613.
53. McKay, G., Otterburn, M.S. and Sweeny, A.G. (1981). Surface mass transfer processes during colour removal from effluent using silica. *Water Res.*, 15, 327–331.
54. Zhu, M.X., Xie, M. and Jiang, X. (2006). Interaction of fluoride with hydroxyaluminum-montmorillonite complexes and implications for fluoride-contaminated acidic soils. *Appl. Geochem.*, 21, 675–683.

55. Chen, J.P. and Wang, L. (2004). Characterization of metal adsorption kinetic properties in batch and fixed-bed reactors. *Chemosphere*, 54, 397–404.
56. Kannan, N. and Meenakshisundaram, M. (2002). Adsorption of Congo red on various activated carbons. A comparative study. *Water Air Soil Pollut.*, 138, 289–305.
57. Acemioglu, B. (2005). Batch kinetic study of sorption of methylene blue by perlite. *Chem. Eng. J.*, 106, 73–81.
58. Pavlatou, A. and Polyzopoulos, N.A. (1988). The role of diffusion in the kinetics of phosphate desorption: The relevance of the Elovich equation. *Eur. J. Soil Sci.*, 39, 425–436.
59. Chen, J.P. and Lin, M.S. (2001). Equilibrium and kinetic of metal ion adsorption onto a commercial h-type granular activated carbon: Experimental and modeling studies. *Water Res.*, 35, 2385–2394.
60. Stollenwerk, K.G. (2003). Geochemical processes controlling transport of arsenic in ground water: A review of adsorption. In: Welch, A. H. and Stollenwerk, K. G., (Eds.), *Arsenic in Ground Water*, pp. 67–100. Boston, MA: Kluwer Academic Publishers.
61. Sposito, G. (1984). *The Surface Chemistry of Soils*. New York: Oxford University Press.
62. Richter, E., Wilfried, S. and Myers, A.L. (1989). Effeect of adsorption equation on prediction of multicomponent adsorption equilibria by the ideal adsorbed solution theory. *Chem. Eng. Sci.*, 44, 1609–1616.
63. Genc-Fuhrman, H., Tjell, J.C. and Mcconchie, D. (2004). Adsorption of arsenic from water using activated neutralized red mud. *Environ. Sci. Technol.*, 38, 2428–2434.
64. Altundogan, H.S., Altundogan, S., Tumen, F. and Bildik, M. (2000). Arsenic removal from aqueous solutions by adsorption on red mud. *Waste Manage.*, 20, 761–767.
65. Thurman, E.M. (1985). *Organic Geochemistry of Natural Waters*, p. 497. Dordrecht, the Netherlands: Kluwer Publishers.
66. Fan, L., Harris, J.L., Roddick, F.A. and Booker, N.A. (2001). Influence of the characteristics of natural organic matter on the fouling of microfiltration membranes. *Water Res.*, 35, 4455–4463.
67. Xu, H., Allard, B. and Grimvall, A. (1991). Effects of acidification and natural organic matter on the mobility of arsenic in the environment. *Water Air Soil Pollut.*, 57, 269–278.
68. Warwick, P., Inam, E. and Evans, N. (2005). Arsenic's interaction with humic Acid. *Environ. Chem.*, 2, 119–124.
69. Zhang, Y., Yang, M. and Huang, X. (2003). Arsenic(V) removal with a Ce(IV)-doped iron oxide adsorbent. *Chemosphere*, 51, 945–952.
70. Courtijn, E., Vandecasteele, C. and Dams, R. (1990). Speciation of aluminum in surface water. *Sci. Total Environ.*, 90, 191–202.
71. Redman, A.D., Macalady, D.L. and Ahmann, D. (2002). Natural organic matter affects arsenic speciation and sorption onto hematite. *Environ. Sci. Technol.*, 36, 2889–2896.
72. Moreno-Castilla, C. (2004). Adsorption of organic molecules from aqueous solutions on carbon materials. *Carbon*, 42, 83–94.
73. Benefield, D.L., Judkins, F.J. and Weand, L.B. (1982). *Process Chemistry for Water and Waste Water Treatment*. Englewood Cliffs, NJ: Prentice-Hall, Inc.

Index

A

AA, *see* Activated alumina
Activated alumina (AA), 101–103
Acute symptoms, 79
Adsorbent alumina cement granules (ALC)
  Arrhenius equation, 155–156
  batch studies, 125–128
  best-fitting isotherm model, 137
  Bohart and Adams model, 147
  capacity, 128–129
  characterization of, 124–125, 148–149
  Clark model, 146
  columns, behavior of, 141–143
  concentration and dose variation studies, 138
  contact time, 139
  desorption properties of, 177–179
  Dubinin–Radushkevich isotherm models, 136–137
  Elovich equation, 154–155
  Freundlich isotherm models, 135–136
  Hutchins BDST model, 143–144
  intra particle surface diffusion model, 154
  ionic strength, 141
  isotherm studies, 162–163
  kinetic modeling, *see* Kinetic modeling
  Langmuir isotherm models, 134–135
  in natural and synthetic systems, evaluation of, 163–167
  in natural groundwater, 171–175
  natural water, 176–177
  pH and coexisting ions, effect of, 139–140
  pseudo-first-order model, 153
  pseudo-second-order model, 153–154
  rate-limiting step, elucidation of, 132–133
  reagents and, 123
  regeneration of, 147–148
  synthesis of, 124
  synthetic water, 175–176
  temperature, 140–141
  Thomas model, 144–145
  Yoon–Nelson model, 145
Adsorbents, 96–99
  dosages, 138, 150
Adsorption
  activated alumina, 101–103
  alumina, 99–101
  alumina-based adsorbents, 103–106
  bone and bone charcoal, 93–94
  carbonaceous and other adsorbents, 96–99
  clays and soils, 94–96
  energy, 134
  rate, 139, 140, 149
  zone, 141
Affinity parameter, 135
African scenario, *see* Asian scenario
Agency for Toxic Substances and Disease Registry (ATSDR), 45–46
Agitation rate, fluoride removal, 139, 149–151
Airborne fluorides
  environmental transport, distribution, and transformation, 60–61
  inhalation of, 62
Alkaline soda water, 7
Alum, 90–92
Alumina, 99–103
Alumina-based adsorbents, 103–106
Alumina cement granules (ALC), 123
  adsorbent, *see* Adsorbent alumina cement granules (ALC)
  composition of, 124
  in natural and synthetic systems, evaluation of, 163–167
  in natural groundwater, 171–175
  x-ray diffractogram of, 149
Amelogenins, of dental fluorosis, 31
Ando soils, 94

Animals, fluoride effect, 78–79
Aquatic organisms, fluoride effect, 76–77
Arrhenius equation, 132, 155–156
Asian scenario, 11–13
 intensity and severity of excess fluoride, 14–18
 South Africa, 20
 Tanzania, 19
ATSDR, *see* Agency for Toxic Substances and Disease Registry

B

Batch studies, fluoride removal
 column studies, 127–128
 equilibrium studies, 126–127
 process parameters, effects of, 125–126
BDST model, *see* Bed depth service time
Bed depth service time (BDST) model, 143–144, 183
 in synthetic water, 171
BET method, *see* Brunauer, Emmett and Teller method
Beverage, fluoride from, 63–68
Bipolar electrode system, 108
Bohart and Adams model, 147, 175, 176
Bone, 93–94
Bone charcoal, 93–94
Breakthrough curve, process parameters effects on, 168–171
Bronsted acids, 99–100
Brunauer, Emmett and Teller (BET) method, 124
Bufo melanostictus, 76
Bulk density, 125

C

Calcareous minerals, 6
Calcium compounds, 89
Calcium fluoride, 60
Cancer, 51–53
Carbonaceous, 96–99
Carbon nanotubes (CNTs), 103
Cardiovascular system, 55
Cariostatic activity of fluoride, 28
CDC, *see* Centers for Disease Control and Prevention
Centers for Disease Control and Prevention (CDC), 52
Cerebella cortex, 54
Chemical profile, fluorine, 3
Chemisorption, 93, 99
China, fluoride map of, 12
Chitin, 98
Chronic childhood diseases, 28
Clark model, 146, 174, 176
Clays, 94–96
CNTs, *see* Carbon nanotubes
Coagulation process, 87–89
Column studies, fluoride removal, 127–128, 167–171
Concentration variation studies, fluoride removal, 138
 in synthetic water, 126
Crippling skeletal fluorosis, 43–44, 45
CRS, *see* Cryolite recovery sludge
Cryolite recovery sludge (CRS), 76
Crystal lattice stability, of enamel, 28

D

DDUs, *see* Domestic defluoridation units
Defluoridation technique, 114–115
 adsorption, *see* adsorption
 coagulation, 87–89
 co precipitation of fluoride, 90–92
 electrochemical methods, 106–109
 membrane processes, 109–114
De ionized (DI) water, 123
Dental caries, 28–29
 by fluoride, prevention of, 29
 role of fluoride in, 29–30
Dental effects, of fluoride, 27–30
Dental fluorosis, 19–20
 dental effects of fluoride, 27–30
 development of, 31–32
 history and occurrence, 30
 issues of, 33
 physical symptoms, 32
 prevalence of, 33–35
Dental plaque, 28, 29
Dental products, fluoride from, 62–63
Dental saliva, 28, 29
Dental treatments, 28

Desorption studies, fluoride, 177–179
Diffusion control process model, 132, 154
Dilution method, 127
DI water, *see* De ionized water
Domestic defluoridation units (DDUs), 89, 102, 142, 176
Dose variation studies, fluoride removal, 138
Draw type, of defluoridation system, 92
Drinking water, fluoride in
  geogenic pollutants, 3–4
  in groundwater, 5–7
  scenario, 1–2
D–R isotherm models, *see* Dubinin–Radushkevich isotherm models
Dubinin–Radushkevich (D–R) isotherm models, 136–137

E

EC, *see* Electrocoagulation
ED, *see* Electrodialysis
EDX, *see* Energy-dispersive x-ray
Electrochemical methods
  electrocoagulation, 106–108
  electrosorption, 109
Electrocoagulation (EC), 106–108
Electro-condensation effect, 107
Electrodialysis (ED), 113–114
Electrosorption, 109
Elemental fluorine, 3
Elovich equation, 131–132, 154–155
Elucidation of rate-limiting step, 132–133, 156–158
Enamel maturation, of dental fluorosis, 31
Endemic skeletal fluorosis, 41
Energy-dispersive x-ray (EDX), 124
Environmental distribution, of airborne fluorides, 60–61
Environmental exposure, sources of, 59–60
Environmental levels, *see* Human exposure
Environmental transformation, of airborne fluorides, 60–61
Environmental transportation, of airborne fluorides, 60–61
Equilibrium sorption studies, fluoride removal, 126–127
EXAFS, *see* Extended x-ray absorption fine structure
Extended x-ray absorption fine structure (EXAFS), 141

F

FAAS, *see* Flame atomic absorption spectrophotometer
Fill type, of defluoridation system, 92
Flame atomic absorption spectrophotometer (FAAS), 124
Fluoridated mouth rinses, 28, 29
Fluoride aids, 29
Fluoride, co precipitation of
  alum, 90–91
  lime, 91–92
Fluoride map
  China, 12
  India, 13
  South Africa, 20
  Tanzania, 19
Fluoride pollution scenario, 11
  global scenario, *see* Global scenario
Fluoride sorptive removal, mechanisms of, 158–162
Fluoride toothpastes, population, 28
Fluorosilicic acid, 60
Food, fluoride from, 63–68
Fourier transform infrared (FTIR) analysis, 125
Freundlich isotherm models, 135–136, 146, 174
FTIR analysis, *see* Fourier transform infrared analysis

G

Gaseous fluorides, 60
Gastrointestinal system, fluoride and, 53–54
Gelatin, 104
Genu valgum, 43
Genu varum, 43
Geogenic pollutants, 3–4, 13, 21

Geological process, 21
Gibbs free energy, 140
Global fluoride map, 12
Global scenario
  China, 12
  India, 13
  Indian scenario, 20–21
  intensity and severity of excess fluoride, 14–18
  South Africa, 20
  Tanzania, 19
Glutaraldehyde (GTA), 104
Granite rocks, 4
Green Revolution, 2
GTA, *see* Glutaraldehyde
Guidelines for Drinking Water Quality, 79

H

HDPE, *see* High-density polyethylene
Health effects, 54–55
Heilongjiang province, 55
HF, *see* Hydrogen fluoride
HFO, *see* Hydrous ferric oxide
High-density polyethylene (HDPE), 114
Human bones, fluoride on, 39–40
Human exposure, 61–62
  dental products, 62–63
  fluoride in soil, 68–69
  food and beverage, 63–68
  occupational exposure, 69–71
  TF exposure, 71
Humans milk, fluoride in, 5
Humans, stress effects of fluoride on
  cancer, 51–53
  gastrointestinal system, 53–54
  health effects, 54–55
Humic acid solutions, effects of, 127
Hutchin bed depth service time (BDST) model, 143–144
Hydro-fluoro-aluminum, concept of, 108
Hydrogen fluoride, 60
Hydrogen fluoride (HF), 5
Hydrogen-ion concentration, 6
Hydrolytic reactions, 90
Hydrous ferric oxide (HFO), 97
Hydroxyl groups, 95, 99–100

I

IEP, *see* Isoelectric point
India, fluoride map of, 13
Indian Medical Gazette, 30
Indian research revelations, 45
Indian scenario, 20–21
Infectious disease, dental caries, 28–29
Intellectual ability (IQ), for children, 54–55
Intra particle surface diffusion model, 131, 154
Ion-exchange mechanism, 7
Ion-exchange resins, 106
Ionic strength, 141
  on fluoride, effect of, 164–166
Ions, effects of, 139–140, 166–167
Isoelectric point (IEP), 103
Isotherm models, 162–163

K

Kinetic modeling
  Arrhenius equation, 132, 155–156
  Elovich equation, 131–132, 154–155
  intra particle diffusion model, 131, 154
  pseudo-first-order model, 129–130, 153
  pseudo-second-order model, 130–131, 153–154
Kinetics studies, fluoride removal
  adsorbent dosage, 150–151
  agitation rate, 149–150
Knozinger's model, 99
Kyphosis, 43

L

Laboratory animals
  fluoride exposure on, 72–75
  in vitro systems, fluoride effects on, 71
Lactobacilli, in dental plaque, 29
Langmuir isotherm models, 134–135
Layered double hydroxides (LDHs), 98–99
LDHs, *see* Layered double hydroxides
Ligand-exchange model, 101, 159, 161

Lime, 88, 91–92
Linear low-density polyethylene (LLDPE), 114
Linear plot, 131
Linear regression, of Thomas model, 173, 183
LLDPE, *see* Linear low-density polyethylene
Long-term air-monitoring, 70
Low-pressure RO membranes, 112
Low-solubility product, 6, 7

M

Magnesium oxide, 88–89
Malignant bone tumor, 52
Material mass balance equation, 142
Membrane processes, 109–110
  electrodialysis, 113–114
  nanofiltration, 112–113
  reverse osmosis, 110–112
Meta-analyses, 54
Mild dental fluorosis, 31
Mottled enamel, dental fluorosis, 30
Mugil cephalus, 76
Multifactorial disease, dental caries, 28–29
Muscovites, 7

N

Nalgonda, 21
Nalgonda technique, 91–92
Nanofiltration (NF), 112–113
National Institute for Occupational Safety and Health (NIOSH), 70
National Occupational Exposure Survey (NOES), 70
Natural groundwater
  ALC in, 171–175
  characteristics of, 168
Natural organic matter (NOM), 166–167
Natural water, 176–177
Nephelo turbidity meter, 124
Nexus™ 870 spectometer, 124
NF, *see* Nanofiltration
NIOSH, *see* National Institute for Occupational Safety and Health
NOES, *see* National Occupational Exposure Survey
NOM, *see* Natural organic matter
Nonskeletal fluorosis, 51

O

Occupational exposure, fluoride from, 69–71
Osteosclerosis, 40

P

Palaemon pacificus, 76
Pan masala, fluoride in, 69
Pernaperna, 76
pH measurement
  effect of, 139–140, 164–166
  of synthetic water, 127
Phosphate compounds, 89
Physisorption, 93
Plants, fluoride effect on, 77–78
Plasticware, 123
Poker back, 41
Polymer-based separation membranes, 111
Pore diffusion, 139
Postshift urinary fluoride levels, 71
Preshift urinary fluoride levels, 71
Pseudo-first-order equations model, 129–130, 153
Pseudo-second-order equations model, 130–131, 153–154
Purkinje cells, 54

R

Rajiv Gandhi Drinking Water Mission (RGDWM), 102
Rate-limiting step, elucidation of, 132–133, 156–158
Reaction-controlled process model, 132, 154
Remineralization process, 40
Renal excretion, of fluoride, 5
Renal failure, 43
Reverse osmosis (RO), 110–112
RGDWM, *see* Rajiv Gandhi Drinking Water Mission
RO, *see* Reverse osmosis

S

Salinity, 113
Scanning electron microscopy (SEM), 124
  of alumina cement granules, 150
SEM, *see* Scanning electron microscopy
Serum fluoride, levels of, 80
Skeletal fluorosis
  action of fluoride on bone, 39–40
  crippling skeletal fluorosis, 43–44
  developments of, 46
  fluoride exposure level and skeletal fracture, 40–41
  level of fluoride and effects of, 44–45
  overview of, 41–43
  significance of, 45–46
Skeletal fracture, 40–41
Sodium fluoride, 60
Sodium hexafluorosilicate, 60
Soil–clay complex, 6
Soil, fluoride in, 68–69
Soils, 94–96
Sorption, 95, 114
  activation energy of, 132
  application of, 171–175
South Africa, fluoride map of, 20
Streptococcus mutans, in dental plaque, 29
Subsurface hypomineralization, in dental fluorosis, 31
Sulfur hexafluoride, 60
Surface complex formation model, 159
Synthetic water, 126, 175–176
  BDST model in, 171
  pH of, 127
  Wolborska model for, 174
  Yoon–Nelson model for, 173
Systemic infection, in dental caries, 29

T

Tanzania, fluoride map of, 19
TDS, *see* Total dissolved solid
TF exposure, 71
Thomas model, 144–145, 176
  linear regression of, 173, 183
Tobacco, fluoride in, 69
Tooth decay, *see* Dental caries
Topography, 71
Total dissolved solid (TDS), 114
Tylodiplax blephariskios, 76

U

UF, *see* Ultrafiltration
Ultrafiltration (UF), 87
Unhygienic oral cavities, 29
UNICEF, *see* United Nations Children's Fund
United Nations Children's Fund (UNICEF), 102
United States, annual incidence rate in, 52
Urinary fluoride, 53, 80
Urolithiasis, hospital admission rates for, 53
U.S. National Academy of Sciences Institute of Medicine, 79–80
U.S. Public Health Service (USPHS), 94

W

Water crisis, 2
Water quality issues, 13
WHO, *see* World Health Organization
Wolborska model, 146, 174, 176
World Health Organization (WHO), 1, 52
  assessment, 13
  guidelines values and standards, 79–80
World Oral Health Report, 27

X

X-ray diffraction analysis (XRD), 125
X-ray diffractogram, ALC, 149
XRD, *see* X-ray diffraction analysis

Y

Yoon–Nelson model, 145
  for synthetic water, 173

Z

Zeolite, 104–106